THE

GANNET'S
GASTRONOMIC
MISCELLANY

For Emma

An Hachette UK Company
www.hachette.co.uk

First published in Great Britain in 2017
by Mitchell Beazley, a division of
Octopus Publishing Group Ltd
Carmelite House, 50 Victoria Embankment
London EC4Y 0DZ
www.octopusbooks.co.uk
www.octopusbooksusa.com

Distributed in the US by
Hachette Book Group
1290 Avenue of the Americas
4th and 5th Floors
New York, NY 10104

Distributed in Canada by
Canadian Manda Group
664 Annette St.
Toronto, Ontario, Canada M6S 2C8

ISBN 978-1-78472-399-6

A CIP catalogue record for this book is
available from the British Library.

Printed and bound in China
10 9 8 7 6 5 4 3 2 1

Publishing director Stephanie Jackson
Senior editor Pauline Bache
Copyeditor Jo Smith
Art director Juliette Norsworthy
Cover design Juliette Norsworthy
Book design Untitled
Illustrations David Bray
Production controller Meskerem Berhane

Thanks to the following for permission to use
their coffee art on page 71. Elvis Matiejunas,
@elvis_seivijus_matiejunas, Facebook
@elvislatteart & @latteartdices; Caleb Tiger
Cha, @calebtiger, Facebook Caleb Tiger
Cha; Kohei Matsuno, @latte_stagram; Paulo
Asi, @traettal; Matthew Lakajev, @latteartporn
& @littleboxcoffee, Facebook LittleBoxCoffee;
Ian Huang via Oracle Coffee Company,
Taiwan, @barista_ian_huang &
@oracle7272672; James Hansen,
@jameskhansen. p110 casabuonarotti.it.

THE GANNET'S GASTRONOMIC MISCELLANY

Killian Fox

MITCHELL BEAZLEY

CLOCKWORK DINERS

Some people crave variety in their dining routines; others go to the same place every day and eat exactly the same thing, perhaps so they can stay focused on matters loftier than choosing a restaurant or agonizing over menus. Here are five renowned creative individuals and their highly predictable eating habits.

» Almost every morning for 15 years, the painter **Lucian Freud** had breakfast at Clarke's restaurant in London's Notting Hill – often returning a few hours later for lunch. He would arrive at 7.30am with his assistant David Dawson and consume saucer-sized *pains aux raisins* or Portuguese custard tarts with extra-milky coffees (referred to by staff as "Mr Freud lattes"). After innumerable hours sitting in her restaurant, Freud invited owner Sally Clarke to his Victorian townhouse a few doors along on Kensington Church Street to sit for a portrait. He painted her three times, the final work interrupted – along with a decade and a half of loyal custom – by Freud's death in 2011.

» Five evenings a week, artists **Gilbert & George** walk nearly 2½ miles from their home in Spitalfields, East London, to dine at the Turkish restaurant Mangal 2 on Stoke Newington Road. They were attracted, they say, by the posh tablecloths and testicles (*koç yumurtası*, or "lambs' testicles", a speciality of the house).

» The novelist **Patricia Highsmith** ate the same thing for virtually every meal: bacon and fried eggs. She began each writing session with a stiff drink – "not to perk her up", according to her biographer, Andrew Wilson, "but to reduce her energy levels, which veered towards the manic". Then she would sit on her bed surrounded by cigarettes, coffee, a doughnut and an accompanying saucer of sugar, the intention being "to avoid any sense of discipline and make the act of writing as pleasurable as possible".

» In the 1960s, the pianist **Glenn Gould** ate almost exclusively at a diner named Fran's near his apartment in Toronto, turning up at 2am or 3am and always ordering the same thing: scrambled eggs, salad, toast, juice, sherbet and decaf coffee.

» The Swedish filmmaker **Ingmar Bergman** always ate the same lunch, according to actress Bibi Andersson, who starred in many of his films. "It doesn't change. It's some kind of whipped sour milk, very fat, and strawberry jam, very sweet – a strange kind of baby food he eats with corn flakes."

SWEET PASTRIES OF THE WORLD

…and their key ingredients.

Cannoli (Italy)
Ricotta

Mooncake (China)
Red bean paste

Danish (Denmark)
Butter

Pastel de nata (Portugal)
Custard

Baklava (Turkey)
Syrup or honey

Kolompeh (Iran)
Dates

Joulutortut (Finland)
Prune jam

Éclair (France)
Chocolate/coffee mousse

Conchas (Mexico)
Cookie dough

NATIONAL APPETITES

Finland consumes more coffee per capita than any other country on earth: the average Finn gets through 1,252 cups a year (or 3.43 cups a day). Here are the biggest consumers of 16 other gastronomic goods.

Food/drink	Country	Annual consumption per capita
Meat	Australia	90.2kg/198¾lb
Fish	Maldives	181kg/399lb
Olive oil	Greece	17.9kg/39½lb
Wine	Vatican City[1]	54.3l/115pt
Rice	Myanmar	227kg/500½lb
Onions	Libya	33.6kg/74lb
Potatoes	Belarus	183kg/403½lb
Eggs	Japan	18.9kg/41¾lb
Beer	Czech Republic	143l/302pt
Sugar	USA	46.1kg/101¾lb
Bananas	Uganda	191kg/421lb
Cacao	Switzerland	9kg/19¾lb
Cheese	France	25.9kg/57lb
Milk	Finland	361kg/796lb
Tea	Turkey	3.2kg/7lb
Alcohol	Belarus	17.5l/37pt

[1] *See* page 47 for an investigation of this curious statistic.

RESTAURANTS IN ODD LOCATIONS

Redwoods Treehouse, *New Zealand*	In a tree
Fortezza Medicea, *Italy*	In a prison
New Lucky Restaurant, *India*	In a cemetery
El Diablo, Lanzarote, *Spain*	On an active volcano
Ithaa, *Maldives*	5m (16ft) under the sea
Dinner in the Sky, *Belgium*	Suspended from a crane
Miners' Tavern, *Poland*	125m (410ft) down a salt mine

INGENIOUS DISHES OF THE 21ST CENTURY

Molecular gastronomy, a term rejected by almost everyone it's applied to, refers to a scientific approach to food preparation and is synonymous with the highly technical cooking of chefs like Ferran Adrià and Heston Blumenthal. It also implies a spirit of playfulness: dishes are designed to surprise and delight – and often the thing on your plate is not what you think it is. Consider these four examples.

Green Apple Taffy Balloon – In 2012, Grant Achatz at Alinea in Chicago challenged his chefs to create a dessert that would float. His deputy, Mike Bagale, discovered a way of inflating a flavoured sugar solution with helium gas. The result was drawn to the table on a string of dehydrated apple, whereupon the diner would pop the balloon with their tongue and suck out the helium, with consequences for vocal cords as well as taste buds.

Artichoke Rose – The most labour-intensive dish on Ferran Adrià's 2009 menu at elBulli, in Spain, was a composite rose that looked and tasted like it was made of sliced artichoke leaves but was in fact composed of organic rose petals. The petals, a digestible variety from Ecuador, had to be picked and separated, blanched three times, put in a pressure cooker, unfurled, pressed to dry and finally arranged in three concentric circles on the plate. This last stage alone took up to ten minutes per dish.

Hot & Iced Tea – One of the plainest-looking courses at Heston Blumenthal's Fat Duck in Bray, England, is also one of the most startling: a cup of tea that's both hot and cold. Nothing visible separates the two vertical halves; the magic is achieved by means of gelling agents that keep the hot and cold liquids from mingling. To disguise the gels, the chefs use malic acid to trigger saliva, making the tea feel thinner in the mouth than it actually is.

Bread of the Forest – On the 2013 menu at Narisawa in Tokyo, the bread course turned up at the start of the meal as a piece of uncooked dough. Left to prove during the appetisers – rising out of its glass tube like a time-lapse mushroom – it was then transferred to a pre-heated stone bowl, covered for 12 minutes and served hot.

THOUSAND-YEAR SHELF LIVES

Forget your mature cheddar, aged beef or 18-year-old whiskey. These foods have been around longer than Christianity and are (possibly) still edible, should you be brave enough to take a bite…

3,000 years old – Pots of honey, still perfectly edible, found in Ancient Egyptian tombs. Honey is said to have an eternal shelf life due to its acidity, lack of water and the presence of hydrogen peroxide. The Ancient Egyptians used it as a sweetener, a gift for the gods and an ingredient in embalming fluid.

FOR YOUR MUMMY

4,000 years old – The world's oldest noodles were unearthed during an archaeological dig on China's Yellow River in 2005. The 50-cm (19½-inch)-long yellow strands – made with grains from millet grass rather than wheat flour – were found in a pot that had probably been buried during a catastrophic flood.

Still fresh

5,000 years old – Bog butter unearthed in County Offaly, Ireland, in 2013. Butter from cow's milk was buried in bogs as a means of preservation, protection from theft or perhaps as divine gifts. After thousands of years in the ground, it has the appearance of paraffin wax, a crumbly consistency and the smell of pungent cheese, with gamey, mossy and funky flavour notes.

36,000 years old – During the excavation of a frozen Steppe bison carcass near Fairbanks, Alaska, palaeontologist Dale Guthrie and colleagues stewed and ate neck tissue while prepping the bison for display. The meat, according to Guthrie, was tough and had – unsurprisingly – a strong aroma.

Bury Me

THE FIVE-SECOND RULE

It's a recurring area of fascination in modern science. If food falls on the floor and you pick it up within five seconds, is it safe to eat? According to Jillian Clarke, a University of Illinois intern who set out to answer the question in 2003 – receiving an Ig Nobel prize for her efforts (*see* page 54) – it generally isn't. She dropped cookies and gummy sweets onto tiles inoculated with *E. coli* for five seconds apiece and found that in all cases the food became contaminated. She also discovered that people are more likely to retrieve cookies and gummy sweets from the floor than broccoli and cauliflower – surprise surprise – and that the rule dates back to Genghis Khan, for whom the grace period extended to 12 hours.

Since Clarke's investigation, further research has either confirmed her findings – Dr Ronald Cutler from Queen Mary, University of London, experimented with *E. coli* and pizza in 2013 and got the same results – or complicated them. Professor Anthony Hilton of Aston University in Birmingham agreed that eating food from the floor can never be entirely risk-free, but pointed out that "the nature of the floor surface, the type of food dropped on the floor and the length of time it spends on the floor can all have an impact on the number [of germs] that can transfer". He delivered this news just as a UK survey revealed that, despite whatever risks there may be, 79 per cent of people admitted to eating food that had fallen on the floor.

ANOSMIA

Around 200,000 people in the UK and more than 2 million in the US suffer from anosmia: they have no sense of smell. One in 5,000 are born with the condition, but most develop it through head trauma or illness. Many people with anosmia lose interest in food, because 80 per cent of the flavour of food comes from its smell.

HOW TO BECOME AN #INSTAGRAMFOODSTAR

The Instagram king of the London food scene, **Clerkenwell Boy** – a 30-something Australian named Tim who keeps his full identity secret – has a knack for finding buzz-worthy dishes and making them look irresistible. Here are his top tips for a successful feed.

Find your niche – "Do you love desserts or burgers or healthy foods, Michelin-star restaurants or home cooking? Figure out what your passion is and define that space. Be consistent and stick to it."

Take good photos – "The main thing is to have a consistent style, whether that's super-close-up food-porn or very clean shots with natural daylight. In a dark restaurant, ask for a table by the window."

Edit with care – "Some people use filters. I try to keep it natural, but Instagram has tools for editing your photos. You might want to make it brighter or moodier, sharpen the dish or blur the edges."

Tag judiciously – "Some of the big Instagrammers use lots of hashtags to get more likes. I don't use that many, only ones I think are relevant – a specific location or type of food."

Write good captions – "…if you want to tell a story. Some people's captions are humorous or cheeky, some are educational, some people just use emojis. I try to keep it as concise as possible."

Time your posts – "It's worth thinking about when you post. On Mondays most people want to eat healthily and might prefer to look at salads or smoothie bowls. Whereas on Friday, when people want to treat themselves, burgers and cocktails might work better."

Interact with the community – "By following people who have a similar style or a feed that interests you, you start to build good connections. You can also use Instagram to put a spotlight on specific causes, like I did for the #CookForSyria campaign."

And finally – "Don't do it just for the likes. The most important thing is to have fun and be yourself."

FOOD HOTSPOTS:
THE MUSIC HANGOUT

Dooky Chase's Restaurant, New Orleans – "I went to Dooky Chase's to get something to eat/ The waitress looked at me and said, 'Ray, you sure look beat.'" This is Ray Charles memorializing one of his favourite New Orleans haunts in his 1961 version of Louis Jordan's *Early In The Morning*. Charles wasn't the only jazz legend to frequent this Treme institution: it also attracted the likes of Duke Ellington, Count Basie, Nat King Cole and Sarah Vaughn, who was a big fan of the stuffed crabs. The restaurant opened in 1941 and flourished under Edgar Lawrence "Dooky" Chase Jr, the jazz-playing son of the original owners, and his wife Leah, who is still running the kitchen today at the age of 94. During the Civil Rights Movement, the restaurant was one of the only public places in New Orleans where African-Americans could meet and discuss strategies. More recently, having weathered Hurricane Katrina in 2005, it has hosted presidents George W Bush and Barack Obama, whom Leah reprimanded for attempting to put hot sauce in his gumbo soup. If you're going, try the gumbo, a favourite of Duke Ellington's, or the red beans as ordered by Ray Charles.

Tom's Restaurant, NYC – Though its exterior is forever linked with the TV sitcom *Seinfeld*, this Greek-American diner at 2880 Broadway was immortalized in a song by Suzanne Vega who, in *Tom's Diner* (1982), drops in for a coffee, skims the newspaper and watches the world go by.

Bar Italia, London – A great late-night coffee shop in Soho, run by the Polledri family since 1949, described by Jarvis Cocker of Pulp in the 1995 song *Bar Italia* as a place "around the corner in Soho/ Where other broken people go". It's actually a much more lively hangout than the lyric suggests.

Café Momus, Paris – The setting for Act 2 of Puccini's opera *La Bohème*, where Alcindoro gets landed with an unwelcome bill, this café was a literary hub in the 19th century. It occupied the ground and first floors of a building on rue des Prêtres-St-Germain-l'Auxerrois, now a hotel called Le Relais du Louvre.

BEST IN THE WORLD!

Making definitive statements about the world's best things, be they restaurants, cheeses or cookbooks, is a pretty silly activity, but it can be fun to play along. Drawn from recent polls and individual authorities, this list should not be considered 100 per cent scientific.

Restaurant	Eleven Madison Park, New York, USA
Bar	The Dead Rabbit, New York, USA
Cookbook	*Mastering the Art of French Cooking*, Simone Beck, Louisette Bertholle and Julia Child
Pizza	Pepe in Grani, Caiazzo, Italy
Burger	Superiority Burger[1], New York, USA
Patisserie	Dominique Ansel, New York, USA
Sushi	Sushi Saito, Tokyo, Japan
Steak	Dons de la Nature, Tokyo, Japan
Fish & chips	Kingfisher Fish & Chips, near Plymouth, UK
Whisky	Booker's Rye 13 Year Old
Bar of chocolate	Bonnat Chocolat's Selva Maya
Beer	Westvleteren 12 (Saint Sixtus Abbey)
Cheese	Tingvollost's Kraftkar, a Norwegian blue
Gin	Hernö Old Tom

Sources: The World's 50 Best Restaurants 2017 (1,040 voters); The World's 50 Best Bars 2016 (476 voters); 1,000 Cookbooks (400+ voters); *Where to Eat Pizza*, Phaidon 2016 (1,077 voters); *GQ* magazine 2015; TW50BR 2017; Asia's 50 Best Restaurants 2017 (318 voters); *Elite Traveller* magazine; National Fish & Chip Awards 2017; *Jim Murray's Whisky Bible* 2017; International Chocolate Awards 2016; RateBeer.com 2016 top 20; World Cheese Awards 2016; World Gin Awards 2017 (best contemporary gin).
[1]Rather controversially, the burger in question is vegetarian with a vegan version available.

BOTRYTIZATION

The condition of being infected with a beneficial form of *Botrytis cinerea*, a grey fungus on wine grapes. In the right conditions, "noble rot", as it is known, can yield grapes that produce a fine sweet wine – aszú from Tokaj-Hegyalja in Hungary/Slovakia and Sauternes in France are famous examples.

THE FATHER AND SON WHO ATE EVERYTHING

The Victorian geologist and palaeontologist William Buckland (1784–1856) was famed not just for digging up dinosaurs and casting a light on the far-distant past, but also for his eccentricity and boundless appetite for new experiences. A committed zoophagist (*see* page 189), Buckland was on a mission to eat his way through the entire animal kingdom. Among the many odd things he is said to have ingested, here are some of the oddest.

Hedgehogs	Mice on toast
Roast ostrich	Common mole
Porpoise	Bluebottle[1]
Crocodile steaks	The mummified heart of King
Cooked puppies	Louis XIV of France[2]

His son Frank (1826–80), though not as eminent a scientist as his father (he was more successful as a popularizer of science than a practitioner), was no less committed to the zoophagic cause. His own eating highlights included:

Elephant trunk	Sea slugs
Porpoise head	Earwigs
Giraffe neck	Curassow
Rhinoceros pie	Horse's tongue
Boa constrictor	Panther chops

[1] Buckland said that mole was the worst thing he'd ever eaten until he tried bluebottle.

[2] This contentious fact comes from the autobiography of Augustus Hare: "Talk of strange relics led to mention of the heart of a French King preserved at Nuneham in a silver casket. Dr Buckland, whilst looking at it, exclaimed, 'I have eaten many strange things, but have never eaten the heart of a king before', and, before anyone could hinder him, he had gobbled it up, and the precious relic was lost for ever."

FIREWATERS OF EUROPE

i. Bagaço (Portugal) – Grape-based spirit sold legally at 37–52 per cent ABV (alcohol by volume), but more potent if homemade. Drunk neat or added to espresso to make *café com cheirinho* – "coffee with a little scent".

ii. Aguardiente (Spain) – Spain's firewater is usually distilled from grapes (in which case it's called *orujo*) and contains over 50 per cent alcohol, though herbs and coffee can also be added to the mix.

iii. Poitín (Ireland) – Traditionally distilled from potatoes in a little pot, hence the name. Outlawed for centuries, it was finally legalized in 1997, though at much lower ABV. Said to curdle milk if unsafe.

iv. Filu 'e Ferru (Sardinia) – A colourless spirit distilled from grape skins. The name ("iron wire") refers to the marker that illicit distillers would supposedly use to mark their underground stashes.

v. Hjemmebrent (Norway) – Means "home burned", though there's a version called *ni seks* ("nine six") which refers to its 96 per cent ABV. Usually potato based. Often mixed with coffee or apple juice.

vi. Pálinka (Hungary) – Any Hungarian can bring fermented mash to a professional who will legally distil this fruit-based spirit (plum, apricot, strawberry and so on), though many still do it at home.

vii. Pontikka (Finland) – This homemade vodka has been banned since 1866, though it still gets made in small quantities. Kitee in the east is known as "the moonshine city of Finland".

viii. Šmakovka (Latvia) – Just one of many words for moonshine in Latvia (others translate as "women's tears" and "the sobs of the ditches"). Made from ingredients such as potatoes, corn and peas.

ix. Šljivovica (Serbia) – Plums are plentiful in Serbia (450,000 tons produced annually) and 70 per cent of them end up in plum brandy, the national drink. Most villages around the country have private stills.

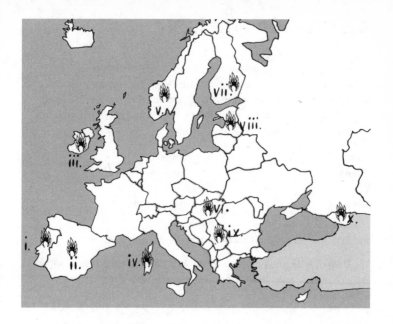

x. Chacha (Georgia) – Wine is big in Georgia, so it makes sense that its (legal) moonshine is mostly grape based, though chacha can also be made with figs, oranges, tangerines or mulberries.

CHICKEN-OF-THE-WOODS

The mushroom *Laetiporus sulphureus* – better known as chicken-of-the-woods – isn't easy to find in shops and restaurants, but hard to miss if you wander past one in the forest (in Europe or North America in late summer or autumn). Vibrant yellow-orange, it clings to the sides of trees in a shelf-like arrangement. Some people are allergic, and it shouldn't be eaten raw, but fry it up and your bravery will be rewarded with a striking meaty flavour, tinged with lemon, that accounts for the mushroom's unusual name.

TASTES LIKE CHICKEN

Some years ago, an American academic named Joe Staton attempted to explain why many exotic meats taste like *Gallus gallus domesticus*, the domestic chicken. He proposed an answer – all chicken-flavoured creatures share a common tetrapod ancestor – and listed a number of meats that fall into this category, including:

Alligator	Snake
Iguana	Snapping turtle
Rabbit	Bullfrog
Kangaroo	Two-toed amphiuma

Should you need inspiration to pursue the enquiry further…

» Chef Stephen Stryjewski of Cochon in New Orleans fries **alligator** tenderloin in a buttermilk batter and serves it with chilli garlic mayonnaise.

» Riffing on the Chinese-American classic General Tso's chicken, blogger Jamie Carlson marinates **snapping turtle** in soy and cornstarch and stir-fries it with chilli, ginger and onion.

» Matt Fitton at the Playford Hotel, Adelaide, serves **kangaroo** tartare with crème fraîche, coal oil and native leaves.

» In *The Lady's Assistant for Regulating and Supplying Her Table* (1787), English writer Charlotte Mason includes a recipe for **viper broth**. "Take a large fowl, draw it, take out all the fat and the breast-bone, fill the body with parsley, a handful of pimpernel, and a head of endive; put these into three pints of Water, with a little salt and pepper; stew it on a slow fire, and let it instill till there is only a quart left; then kill a viper, skin it and take out the entrails, cut the flesh into small pieces, put it with the broth, with the heart and liver cut across, two blades of mace, and a bit of cinnamon; cover it up and let it boil till it is reduced to a pint; by this time the flesh of the viper will be consumed then, strain it off and press it very hard. It will serve twice."

ONE FOR THE POT

...DOES NOT TASTE LIKE CHICKEN

Not all exotic meats call to mind *Gallus gallus domesticus*, however. Here are seven seldom-eaten creatures and the familiar foods to which they've been compared.

Butterflies – In an interview, author and keen lepidopterist Vladimir Nabokov confessed to eating some of his catches in Vermont. "I didn't see any difference between the monarch butterfly and the viceroy," he said. "The taste of both was vile… They tasted like almonds and perhaps a green cheese combination."

Bat – *Atlas Obscura* reporter Lynn Freehill-Maye tried curried fruit bat at a restaurant in the Seychelles, where it is considered a delicacy, and likened it to extra-tough duck.

Muskrat – One participant at the annual Muskrat Dinner in Michigan, USA, compared the cooked flesh of this semi-aquatic rodent to strong, dark turkey meat.

Puma – During his voyage on the *Beagle*, Charles Darwin sampled iguanas, armadillos and this American big cat, which, he reported, was "remarkably like veal in taste".

Termites – Persuaded to eat live termites from a tree trunk in a Nicaraguan rainforest several years ago, I was strongly reminded of mushrooms.

Porpoise head – The Victorian naturalist Frank Buckland made it his mission to eat as many weird things as possible (*see* page 13). In his view, the flesh of a porpoise head tasted like "broiled lamp wick", which raises another question: how did he know what broiled lamp wick tasted like?

Toad tadpoles – For a 1971 paper testing the comparative palatability of eight species of tadpoles, US biologist Richard Wassersug gave a particularly negative review to the larval offspring of a toad. "It was one of the worst things I'd tasted," he says. "So astonishingly bitter. A teaspoon full of Tabasco sauce might get you close."

CULINARY TIME TRAVEL: 18ᵀᴴ-CENTURY BRITAIN

Annie Gray, author of *The Greedy Queen: Eating With Victoria*, is an historian and broadcaster who explores the history of food and dining in Britain from 1600 to the present day. Here, she time-travels to the era in British culinary history that fascinates her most, to attend a riotous dinner party.

The period

"I'll choose the late 18ᵗʰ century. At that point British cuisine is developing a clear identity. You've got most of the ingredients of the modern world in place, things like chilli and chocolate and tea, but it's far enough removed from the modern day that it'd be a real joy to the palate, but close enough that it wouldn't be weird. You're talking rich people's cuisine, obviously."

For starters

"I'd be having a three-course meal with cheese. I would start with soup and fish and then fancy 'made' dishes[1] on my table, probably four or five, and I would try one of each but only in small amounts. I might have mutton chops with a sauce, or veal stew or potato pudding, or maybe a pie, or a small slow-cooked joint."

The second course

"Then all of that would be taken away and the handsome footman would lay out the table for my second course. Being a woman, I'd be at the end that has the chicken on it. The man at the other end would have hare in front of him, but it's alright because I can pick and choose what I want. There'd be a lot of roast meats – very lovely, all spit-roasted rather than baked in an oven. On the table as well would be vegetables, probably cardoons, Jerusalem artichokes, asparagus, peas. And the sweet dishes would be on the table at the same time: plum pudding, a mark of Britishness, and maybe some blancmange or jelly."

Cheese & dessert

"Then all of that goes away and in comes my cheese course. After that would be dessert, which is just palate cleansing, so I'd look for hothouse fruit, probably a pineapple[2], on the table. If my host is rich enough he probably has hothouses and there might be nectarines too (if you're rich enough, you don't eat in season). And there'd probably be some ice creams on the table as well, water ices, hopefully a nice boozy sorbet."

To drink

"I should mention that everyone's really trolleyed because there's lots of booze. Wine if you're rich enough, beer if you're not. The gents will be sitting there getting buggered on punch and missing the chamber pot which is concealed behind the sideboard – there are pictures: we all know it happens."

[1] A dish of food prepared from several ingredients (such as meat, vegetables and herbs).
[2] Annie: "It's said that if a man had four horses, they produced enough manure [to help to grow] two pineapples every month of the year."

ENTOMOPHAGY, OR EATING INSECTS

Insects are among the most efficient forms of food available: 80 per cent of a cricket can be eaten, while only 55 per cent of a pig and 40 per cent of a cow are edible.

Crickets also require less water, land, pesticides and 12 times less feed than cattle, while providing a competitive amount of protein – 8–25g per 100g (¼–1oz per 3½oz) compared to 19–26g per 100g (¾–1oz per 3½oz) of raw beef. They emit much smaller amounts of greenhouse gases too: 80 times less methane than cows and 8–12 times less ammonia than pigs.

MORE BUG STATS

1,900 – Number of edible species of insect
2 billion – People eating insects on a regular basis
1.4 billion – Insects to every human on earth
300g (10½oz) – Average weekly household consumption of caterpillars in Kinshasa, Congo
1kg (2lb 4oz) – Amount of insects unwittingly consumed by every person on earth, including the 5 billion non-insect-eaters, each year

WHERE TO START

Grasshoppers – In Mexico, they are fried until crispy and served with chilli and lime. I've tried them: they're delicious.
Ants – Roasted with salt and eaten at feasts in Colombia. Live black ants were famously served at Noma in Copenhagen.
Worms – Earthworms, buffalo worms and mealworms are all edible. Maguey worms don't just end up in tequila bottles: they are also fried and apparently taste a little like sunflower seeds.

GREAT MEALS IN LITERATURE: BREAKFAST

When we first meet Dublin ad man Leopold Bloom in chapter 4 of **Ulysses** (1922) by James Joyce, we are informed that he "ate with relish the inner organs of beasts and fowls". Joyce elaborates in exquisite detail:

> He liked thick giblet soup, nutty gizzards, a stuffed roast heart, liver slices fried with crustcrumbs, fried hencod's roes. Most of all he liked grilled mutton kidneys which gave to his palate a fine tang of faintly scented urine.

It's kidneys that Bloom has in his mind as he prepares to have breakfast on 16 June 1904, the epic day on which the novel unfolds. As it's a Thursday, however – "not a good day either for a mutton kidney over at Buckley's" – he decides to buy a pork kidney instead.

Leaving the house, Bloom walks to Dlugacz's butchers where, in the window, "A kidney oozed bloodgouts on the willowpatterned dish: the last." It costs him threepence. He slides the "moist tender gland" into a side pocket, takes it home and drops it into the hot pan, letting the bloodsmeared paper fall to the cat. Then he gets distracted by his wife Molly – who wonders, moments later, if she smells something burning in the kitchen…

> Pungent smoke shot up in an angry jet from a side of the pan. By prodding a prong of the fork under the kidney he detached it and turned it turtle on its back. Only a little burnt. He tossed it off the pan on to a plate and let the scanty brown gravy trickle over it. Cup of tea now. He sat down, cut and buttered a slice of the loaf. He shore away the burnt flesh and flung it to the cat. Then he put a forkful into his mouth, chewing with discernment the toothsome pliant meat. Done to a turn.

See also…
» Patrick Bateman's bran muffin breakfast in *American Psycho* (1991) by Bret Easton Ellis.

THE GREAT WINE FORGERY

In 2006, a record $24.7m of vintage wine from a single cellar was sold at auction in New York. The owner of the cellar was one **Rudy Kurniawan**, a 29-year-old Indonesian who had cropped up on the LA wine scene in the early 2000s boasting an excellent palate and seemingly limitless cash to spend on the world's most sought-after bottles (his thirst for Domaine de la Romanée-Conti wines earned him the nickname "Dr Conti"). When he began selling his stock, other collectors were astounded by his steady supply of ultra-rare vintages (he sold the billionaire Bill Koch a jeroboam of Château Mouton-Rothschild 1945, later revealed to be a fake, for $48,259). It took a while before anyone began to question him publicly: one key challenger was the Burgundy winemaker Laurent Ponsot, who discovered Kurniawan was selling vintages of Domaine Ponsot wines that had never been produced.

Eventually, in March 2012, the FBI raided Kurniawan's house in Arcadia, Los Angeles County, and found fake labels, corking tools and empty bottles soaking in a sink. Using extensive tasting notes, he had been mixing cheaper (but still good) wines to approximate much more expensive bottles. Sentenced to ten years in prison, Kurniawan was the first person ever to be convicted of wine fraud, though far from the first to deceive a bunch of wine connoisseurs with inferior product.

OTHER GASTRONOMIC IMPOSTERS

Red snapper – A 2013 study found that 38 per cent of all US restaurants made false claims about the fish on the menu. Red snapper is one of the most mislabelled. "Just don't ever order red snapper," advised one expert, "in more than 94 per cent of cases, you're not getting the real thing."

Olive oil – Many olive oils claiming to come from Italy merely passed through that country en route to your cupboard. In 2015, seven major Italian companies (including Carapelli and Bertolli) were accused of passing off inferior olive oil as "extra virgin". In other cases, olive oils were found to have been diluted with vegetable oil and coloured with beta-carotene.

Kobe beef – Be wary of steak joints claiming their beef comes from the Tajima strain of Wagyu cattle, raised in the Hyogo Prefecture in Japan under rules set out by the Kobe Beef Marketing Association. Only 78 restaurants outside Japan (20 in the UK and US combined) are licensed to buy and serve this hallowed meat.

Caviar – While poaching of endangered sturgeons is the bigger concern, fraud is a major problem in the caviar trade. Mislabelling is rife and tests have exposed blatant counterfeits: you can make a pretty convincing fake caviar using vegetable oil, flavoured liquid, gelatin powder and a pipette.

Saffron – Laborious to harvest and relatively easy to imitate (in appearance if not in flavour), it's no wonder this lavishly expensive spice is regularly faked – dyed stamens from normal crocuses are common culprits. To test, briefly suck on a few strands, spit into a clean tissue and rub: a yellow stain is a positive sign, red is not.

KEEPING IT IN THE FAMILY

Each of the following food-related businesses have been overseen by the same family for more than ten generations.

Est.	Name	Business & location
718	Hoshi Ryukan	Hotel, *Kamatsu, Japan*
1141	Barone Ricasoli	Wine & olive oil, *Siena, Italy*
1304	Hotel Pilgrim Haus	Hotel, *Soest, Germany*
1385	Antinori	Winemakers, *Florence, Italy*
1595	J Epping of Pippsvadr	Grocers, *Pippsvadr, Germany*
1500s	Toraya	Confectioners, *Tokyo, Japan*
1602	Ensho Sadu	Tea school, *Tokyo, Japan*
1630	Kikkoman	Soy sauce, *Noda, Japan*
1637	Gekkeikan	Sake, *Fushimi, Japan*
1642	Barker's Farm	Dairy & apples, *Mass., USA*
1664	Schwartze & Schlichte	Distillery, *Oelde, Germany*

REASSURINGLY EXPENSIVE?

What compels restaurants and food companies to sell certain products for prices high enough to induce altitude sickness? Factors such as extreme rarity, production time and social value go just part of the way towards explaining it. The primary motivation is usually the barrage of media coverage – followed by mentions in books such as this – that attends each record-breaking event.

Wine – A 12-bottle case of 1988 Romanée-Conti fetched CHF144,000 (£110,000/$148,000) at an auction in Geneva in 2016. That works out at £2,138/$2,877 per glass.

Lobster – At a 2012 auction in Gothenburg, to mark the start of Sweden's lobster season, the opening sale fetched SEK102,000 (more than £9,000/$12,500) per kilo (2¼lb).

Beef – A rib steak from Boucherie Polmard in Paris, using meat from Blonde d'Aquitaine cows which has been aged for 15 years or more by a process of "hibernation", can cost up to €3,000 (£2,500/$3,500).

Coffee – Geisha coffee beans from Hacienda La Esmeralda in Panama sold for a record $350.25 (£235) per pound (450g) at auction in 2013.

Truffle – In 2010, Macau billionaire Stanley Ho bought an Alba white truffle (the second most expensive thing you can eat, after gold leaf) for $417,200 (£269,000) – or $320.92 (£207) per gram (0.04oz).

Caviar – Strottarga Bianco caviar, made in Austria by mixing the white fish roe of the rare albino sturgeon with 22-carat gold leaf, retails at £200,000 ($260,690) per kilo (2¼lb).

Chocolate – A 50-gram (1¾-oz) bar of Ecuadorian chocolate aged for 18 months in a 50-year-old oak Cognac cask, by a company called To'ak, retails at $345 (£266). The makers deem their product so precious that they include a pair of wooden tongs in the box so you can avoid contaminating the chocolate with your finger odours.

Sandwich – Serendipity 3, a New York restaurant, sells a "Quintessential Grilled Cheese" sandwich – Caciocavallo Podolico cheese, white truffle butter, and bread made with Dom Pérignon Champagne and 24-carat gold flakes – for $214 (£165).

Tea – Ultra-rare Da Hong Pao tea from China's Fujian province goes for $1,400 (£1,080) per gram (0.04oz). The Royal China Club in London serves a four-cup pot for £180 ($235).

Melon – In Japan, a pair of Yubari melons, a type of premium cantaloupe, sold to a supermarket for a record ¥3m (£21,500/$27,140) in the first auction of the 2016 harvest.

Burger – For a mere €2,000 (£1,785/$2,360), chef Diego Buik at South of Houston restaurant in the Hague will make you a Wagyu beef burger with gin-infused lobster, foie gras, caviar, truffle, and a brioche bun covered in 24-carat gold leaf.

D

DOLSOT

Most tableware items have multiple functions. The Korean *dolsot* ("stone pot") has just one: to contain hot rice dishes such as *bibimbap*. The food cooks in the *dolsot*, which is made of agalmatolite, and is carried sizzling to the table, where the diner stirs the elements together and is left, at the end, with a delicious layer of crispy burnt rice at the bottom.

EXTREME LIFESPANS OF EDIBLE THINGS

Of all the living things we eat, the majority are harvested in a matter of months, if planted in the ground, or slaughtered within a year or two if farmed on land or netted at sea. At either end of the spectrum there are a few notable species that live a span of decades – or mere days – before succumbing to the human appetite.

VENERABLE

37 years – The agave plant, from which tequila and mezcal are produced, usually takes around seven years to cultivate, though wild-growing agaves such as *tobolá* don't mature for decades – their slow fruition posing a serious problem for the fast-growing mezcal industry.

507 years – Ming, the oldest-recorded mahogany clam (*Arctica islandica*), was dredged off the coast of Iceland in 2006. These palm-sized bivalves – still delicious after a century or more – are best cut into thin slices and eaten raw to savour their intense marine flavour and subtle crunch.

2,000 + years – The olive tree of Vouves in western Crete dates back at least as far as Christ, though some scientists believe it's up to 4,000 years old. The tree still yields today: 55kg (121lb) of olives were collected during the 2012 harvest.

EPHEMERAL

1–2 weeks – Mushrooms grow freakishly fast, as many time-lapse videos testify. Few edible varieties reach fruition quicker than the oyster mushroom which, if inoculated properly, can go from zero to risotto in under a fortnight.

< 28 days – A poussin (sometimes known as *coquelet*) is a chicken that's slaughtered young, weighing around 400–550g (14oz–1lb 4oz), and usually served in a single portion. Elizabeth David didn't approve: "To me these seem wretched little birds, poor in flavour and stringy in texture."

3 months – The average lifespan of a cricket. A short breeding time is one of many reasons why insect farming is seen as an efficient answer to global food supply problems. All that's needed now is for the 5 billion non-insect-eaters to get over their fear of bugs (*see* page 20).

EATEN TO EXTINCTION

It will surprise no one to learn that more animal species are being
wiped out in this century than at any other time in our history. But
Homo sapiens have always been destructive and the natural world
has been suffering the effects for tens of thousands of years. Human
appetite is at least partly to blame for the following extinctions:

What?	Where?	When?
Woolly mammoth	Asia, Europe, America	4,000 years ago
Meiolania turtle	New Caledonia	3,000 years ago
Eurasian aurochs	Eurasia	1627
Dodo	Mauritius	1681
Steller's sea cow	Bering Sea	1768
Great auk	North Atlantic	1844
Atlas bear	North Africa	1870
Quagga	Southern Africa	1883
Passenger pigeon	North America	1914
Bubal hartebeests	North Africa	1920s
Toolache wallabies	Southern Australia	1937
Caribbean monk seal	Caribbean	2008

ON THE BRINK

What?	Where?
Atlantic bluefin tuna	Atlantic, Mediterranean
South African abalone	South Africa
Atlantic goliath grouper	Atlantic
Pangolin	Africa, Asia
Yellow-spotted river turtle	Amazon Basin
Manatee	Amazon, Caribbean, West Africa
Curassow	Central & South America
Chinese giant salamander	China
Chinook salmon	Pacific northwest
Choctaw hog	North America

THE SHAPES OF FOOD: OBSCURE CUTLERY

The last hundred years or so have thrown up a handful of eccentric eating implements – among them the splayd, a 1940s Australian invention which combines the functions of knife, fork and spoon. But the true pinnacle of hyper-specialist cutlery was the Victorian era in Britain (1837–1901), when service *à la française* (all the dishes coming to the table at once) gave way to service *à la russe* (dinner served in a sequence of courses, each requiring its own utensils).

Lettuce fork
Victorian era, Britain
Dispenses individual lettuce leaves at the table.

Bon bon scoop
Late 1800s, USA
For dispensing sweets in generous quantities.

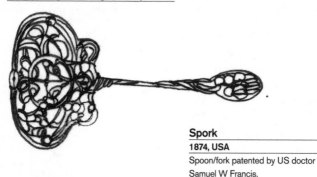

Spork
1874, USA
Spoon/fork patented by US doctor Samuel W Francis.

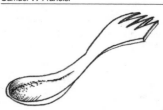

Cheese scoop
18th century, Britain
Traditionally used on Stilton but works for other soft cheeses.

Saratoga chip server
Late 19th century, USA
Named after Saratoga Springs, New York,
home of the potato chip.

Sucket fork
Late 1400s, Britain
Double-ended, for eating sweetmeats,
aka sucket.

Aspic spoon
Victorian era, Britain
Sharp-edged, for cutting through solids
suspended in aspic (savoury jelly).

Bread fork
Victorian era, Britain
For serving bread without touching it
(a cardinal sin at the Victorian table).

Cake breaker
1932, USA
Invented by Cale J Schneider for slicing
angel food cake.

Absinthe spoon
1870s, France
Used to dissolve a sugar cube into a glass
of absinthe.

THE KNIFE WITH A THOUSAND USES

Cutlery complexity is largely a Western phenomenon. On East Asian tables you will rarely find more than chopsticks and soup spoons. Multitasking is evident, too, in Chinese kitchens, where an endlessly adaptable knife, the *tou*, is used for everything from splitting firewood and killing pigs, to crushing garlic and slicing vegetables with laser-like precision.

all THIS AND MORE

DISPATCH

CRUSH

SLICE

SPLIT

SOME PLEASINGLY NAMED PUDDINGS

To my knowledge, no scientific study has ever been made of the relative flamboyancy of dish titles. Were it to happen, though, it wouldn't surprise me to learn that sweet things were on balance more exuberantly named than their savoury counterparts. If you're going to pile ice cream, nuts and brightly coloured summer fruit into a sundae glass and drizzle blood-red syrup all over it, you've no choice but to call it something fabulous like knickerbocker glory.

Baked Alaska – Previously known as "Alaska, Florida" (a reference to its cold-hot extremes?), this meringue-capped ice-cream cake first appeared in print with its current name in *The Original Fannie Farmer 1896 Cookbook*.

Pets-de-nonne – Some sources report that the name is a reference to the heaven-troubling lightness of these cream-filled pastry puffs. Another suggests that a nun passed wind while making one at an Alsatian convent.

Langues-de-chat – A case of name following form: these light, crunchy biscuits (*Katzenzungen* in German, *lingue di gatto* in Italian) look like cats' tongues.

Flummery – *Llymrig* in Welsh means "slippery", much like the texture of this jellied, oat-based pudding.

Singin' hinny – This currant scone whistles as it chars on the griddle, hence "singin'". The Scots dialect "hinny" is affectionate rather than referential, as no honey features in the recipe.

Cherpumple – The layers of cherry, pumpkin and apple pies inside this frosted cake account for the portmanteau name.

Knickerbocker glory – Was this layered cream sundae named after a Manhattan hotel? A New York surname? Some pants? No one is quite certain.

THE GANNET'S
TOP FIVE KITCHEN TOOLS

Over the past few years, *The Gannet* has been visiting interesting people around the world and seeing how they cook and eat at home. We've had the opportunity to rummage around a lot of different kitchens and ask their owners what they couldn't live without. Here are the five most indispensable kitchen utensils according to our interviewees.

1. Knife – Not much work gets done in the kitchen without a dependable blade (and a decent sharpener to go with it). Knives can be aesthetically pleasing as well as practical, hence their position on this list, though you don't need to spend vast sums of money to get a good one.

2. Microplane – These ultra-efficient graters and zesters started life as a woodworking tool by Grace Manufacturing of Arkansas, USA. According to company lore, a Canadian housewife saw the grating potential of their wood rasps in 1994 when she commandeered one from her husband's workshop to make an orange cake. The company modified the rasps and started marketing them to cooks. Now, Microplanes crop up in kitchens everywhere we go.

3. Mandoline – Though extremely useful for shaving everything from vegetables to bottarga, these slicers are usually recommended to us with a warning: watch out for your fingers. The food journalist Marie-Odile Briet, who invited us over for lunch in Paris, has a clever solution: while using her mandoline, she wears a steel-lined butcher's glove.

4. Cocktail shaker – An indispensable piece of kit. As Irish hotelier Justin Green remarked while shaking rhubarb martinis for us at Ballyvolane House in Co Cork (*see* page 58): "Life would be very boring without a drink".

5. Spice grinder – Several people we visited had the smart idea of grinding their spices in a basic coffee mill, rather than pounding at them in a pestle and mortar or buying pre-ground. "I never use spices that are pre-ground, unless it's chilli or turmeric," London-based chef Maud Faussurier tells us.

...AND SOME OF OUR MORE UNUSUAL FINDS

It's not all about utility: we've also come across some very curious kitchen items on our travels. Some do serve a genuine purpose (often working better than a more conventional tool), while others are gloriously inessential and earn their place as a conversation piece.

Swan-feather brush – The food writer and chef Olia Hercules, who grew up in southern Ukraine, has an unusual pastry brush in her London kitchen. "In Ukraine they use feathers – if the dough is very thin, you need something delicate that won't break it," she says. "My dad rescued a swan a couple of years ago and my mum made some of its feathers into a brush. Normally you'd use goose or duck feather but swan feathers work really well."

Saint's tongue jar – "I keep my honey in an 18th-century Irish cut-glass jam jar," says Irish artist Conrad Frankel. "It would have been used for jam or honey. Or, if you went to Italy, you might see one in a church with saints' tongues in it – jars this shape were used to hold the relics of saints. This one makes a lovely clinking noise when you open it. Inside is a sycamore honey dipper and some Greek mountain honey."

Grenade-shaped coffee roaster – "In Israel some years ago," says New Zealand chef Peter Gordon, "I met this genius inventor named Ram Evgi [of Coffee-Tech Engineering]. His roaster resembles a large hand grenade which you fill with green coffee beans and then toast while turning it over an open flame."

Champagne sabre – "A blacksmith friend in San Francisco called Carla made this," says American wine distributor Josh Adler at his apartment in Paris, holding up a large, blunt sword for lopping off champagne bottle-tops. "It's more dramatic than pulling out the cork: you run the sabre up the side of the bottle, along the seam, and the top of the neck comes off clean – in theory at least." Those last words prove fateful: the bottle of sparkling wine that Josh uses to demonstrate his technique explodes when the sabre hits the neck showering the room with glass. Luckily, no one is grievously hurt and the next bottle-top comes off clean.

A BRIEF HISTORY OF KITCHEN TECHNOLOGY

We take our kitchens for granted. But everything that seems so simple to us now, like boiling an egg or brewing a pot of coffee, is based on millennia of slow, incremental developments, many of which came about by pure chance. As food historian Bee Wilson writes in her brilliant book *Consider the Fork*, "The history of food is the history of technology. There is no cooking without fire."

29,000BC – Pit-based ovens are used to roast and boil mammoths in what is now the Czech Republic.

18,000–17,000BC – Earliest-known clay cooking pots developed in southeast China.

500BC – Cast iron is invented in China, allowing for cheaper knives, though the alloy doesn't become available in the West for another 1,000 years.

1100s – Chimneys first appear in northern Europe, allowing much-needed extraction in kitchens, though they don't become common in ordinary houses until the 16th and 17th centuries.

1500s – Mechanized jacks are developed in Britain to turn roasting spits, saving boys, dogs and geese from having to do it manually (*see* page 68).

1637 – In France, Cardinal Richelieu bans sharp, double-bladed knives from his table. Louis XIV later forbids cutlers from forging them. Soon, blunt, single-bladed dinner knives become standard throughout Europe.

1679 – French physicist Denis Papin invents the steam digester, a forerunner of the pressure cooker and the steam engine.

1802 – The first gas stove is developed by Zachaus Winzler in Passau, Germany. It will take 80 years for cooking with gas to catch on.

1813 – The world's first canning factory opens in London. It will be a further 42 years before someone patents a can opener.

1850 – Joel Houghton of New York State patents a proto-dishwasher: a wooden machine with a hand-turned wheel that splashes water on dishes.

1880s – Cooking with gas starts to catch on as manufacturers like William Sugg of London introduce accessible gas ranges.

1913 – Seeking to improve gun barrels, Harry Brearley of Sheffield invents stainless steel and inadvertently revolutionizes the world of cutlery.

1915 – A very handy measuring vessel made of heatproof borosilicate glass, the Pyrex jug is introduced by a New York company.

1922 – Stephen Poplawski, a Polish-American living in Wisconsin, invents the blender. In the same year, Arthur Leslie Large of Birmingham, England, invents the electric kettle.

1938 – Teflon (polytetrafluoroethylene or PTFE) is accidentally discovered by DuPont chemist Roy Plunkett. It is later used in nonstick cookware.

1945 – Percy Spencer of the Raytheon Company invents the microwave while working on military radar systems.

1960s – Sous-vide cooking, whereby food is vacuum-sealed and placed in a water bath at a controlled temperature, is developed as American and French engineers master the technology, though it takes another few decades to gain acceptance in restaurant kitchens.

2005 – Alan Adler, inventor of the Aerobie throwing disc, introduces the AeroPress, an ingenious syringe-like device for plunging coffee through a filter.

A YEAR IN FOOD HOLIDAYS

Whatever day of the year it is, chances are some obscure product is being celebrated in some part of the world via an official food holiday. A few of these have genuine cultural weight – pancakes were traditionally made on Shrove Tuesday in order to use up rich foods before Lent – but a good number of them were invented by marketing execs who simply want to shift more units.

JANUARY 15
Strawberry Day (Japan)
"One" in Japanese is *ichi* and "five" is *go*. *Ichigo* is the Japanese word for strawberry. Go figure.

FEBRUARY 23
Peppermint Patty Day (USA)
It's not clear why a chocolate peppermint confection related to the Kendal mint cake requires an official day, but that day exists nonetheless.

MARCH 30
Idli Day (International)
To mark the inaugural celebration of South India's steamed rice-and-lentil cakes in 2015, a monster 50kg (110lb) idli was produced in the southern city of Chennai.

APRIL 17
Espresso Day (Italy)
Established by the Italian Espresso National Institute in 2008 to promote Italy's black gold. (Not to be confused with USA's National Espresso Day on 23 November.)

MAY 28
Burger Day (International)
Exists alongside a US Cheeseburger Day (18 September) and an entirely separate UK Burger Day (25 August).

JUNE 3
Fish & Chip Day (UK)
It's surprising that such a proud element of Britain's culinary heritage, popular since the mid-19th century, only found its way onto the calendar in 2016.

JULY 21
Lamington Day (Australia)
Not merely a square of sponge rolled in chocolate and desiccated coconut but an Australian institution. Diehard fans have called for this day to be declared a public holiday.

AUGUST Second Sunday
Melon Day (Turkmenistan)
Inspired by Turkmenistan's melon-loving (and dictatorial) first president Saparmurat Niyazov.

SEPTEMBER 5
Cheese Pizza Day (USA)
Just in case you're still hungry after Pizza Day (9 Feb) and can't wait for Pepperoni Pizza Day (20 Sept), Sausage Pizza Day (11 Oct) or Pizza with the Works Except Anchovies Day (12 Nov).

OCTOBER 24
Tripe Day (International)
The official status of this day is uncertain, as the Tripe Marketing Board, which declared it, might not be a 100-per-cent-serious entity. But its motto bears repeating: "Tripe. Yesterday's Food. Today."

NOVEMBER 17
Gose Day (Germany)
Originating in Goslar, Lower Saxony, *gose* is a sour beer with saline undertones, produced by lactic fermentation and often flavoured with coriander.

DECEMBER 4
Cookie Day (USA)
With over 175 days in its calendar dedicated to food holidays, the USA had to squeeze one in somewhere for cookies.

ON ENDOCANNIBALISM

There are two basic types of cannibal society:
Exocannibals are those who eat the flesh of human beings outside their own community, usually their enemies.
Endocannibals are those who eat the flesh of human beings within their own community, usually their relatives.

Whereas eating people is usually associated with hatred and violence, endocannibalism has been viewed (in the few cultures that practised it, such as the Yanomami people of the Amazon rainforest) as an act of respect towards a dead relative, whose life essence might otherwise be lost. According to Canadian scholar Margaret Visser in her eye-opening 1991 book *The Rituals of Dinner*, "A dying endocannibal might catalogue in detail which parts of him or her were later to be eaten by whom: a last will and testament of admirably generous detachment."

MY PERFECT BREAKFAST

This is a particular obsession of mine: I always want to know what people eat for breakfast and how they approach the day's most ritualistic meal. For this reason, there are a lot of reflections on breakfast in *Gannet* interviews. Here are some of my favourites.

"I absolutely love Scandinavian breakfasts: pickled herring, gravadlax, smoked salmon, salami… The best breakfasts I've had were at Hotel Fabian in Helsinki. Every day they have all those things plus fruit compotes, yogurts and *six* different types of rye bread. Honestly, I went to bed every night looking forward to breakfast the next morning." – **Diana Henry, food writer, London, UK**

"I eat the old-school Roman breakfast of *pizza bianca*, a simple flatbread seasoned with salt and olive oil. Before the industrial *cornetti* [Italian-style croissants] trend descended on Italy in the 1970s, Rome's local breakfast was a sweet bun or a slice of pizza, and as an eternal pizza lover I reach for the latter. I follow this up with a double espresso." – **Katie Parla, food writer, Rome, Italy**

"Freshly roasted coffee, ground and brewed with two inches of frothy head. If I'm hungry, I'll eat two (or more) slices of gluten-free brown bread with soft goats' cheese (thickly spread), quartered boiled eggs (cooked just until they are steady on their feet, not hard-boiled) and too many black olives stuffed with fresh ginger. Oh, and a furtive piece of dark chocolate." – **Yemisi Aribisala, food writer, Cape Town, South Africa**

"Depends on the day. I could go a cold navel orange on most of them. But I'm a New Yorker: baconeggncheese on a toasted buttered roll. Eggs over easy. Cut it in half, please, I like the mess." – **Sam Sifton, *New York Times* food editor, New York, USA**

"Coffee, made in an Aeropress, black. Two slices of sourdough toast. Unsalted butter, which must not melt away. On the final half-slice, I might have marmalade. A coffee cup full of wholemilk yogurt sprinkled with pumpkin seeds and a piece of fruit (ideally a blood orange)." – **Bee Wilson, food writer, London, UK**

"A short stack of fluffy pancakes, some crispy bacon and a large glass of orange juice." – **Adrian Miller, soul food scholar, Denver, USA**

"A little porridge and a lot of fresh fruit with some crunchy nuts on top, followed by a slice of buttery toast and a good cup of coffee. All consumed with my nose stuck in a cookbook, planning the next meal." – **Caroline Hennessy, food & drink writer, Cork, Ireland**

"A proper Neapolitan espresso, knocked back at the bar with a chocolate pastry in the other hand; failing that, some good sourdough toast completely saturated with butter, then topped with cream cheese, Marmite and roast tomato would do. I'm easily pleased." – **Thom Eagle, chef & food writer, Suffolk, UK**

"Probably the one served at Coombeshead Farm in Cornwall: bircher muesli made from a variety of ancient grains with a dollop of great yogurt; toast from fresh sourdough with butter and honey on the side; then a carb-light cooked element with belly bacon, hog's pudding and silky scrambled eggs. All eaten in a (relatively) calm room with morning light streaming in, and with four fingers of kombucha and then half a glass of cold apple juice to wake me up." – **Ed Smith, food writer, London, UK**

"I love waffles. I love them most of all when they're heaped with berries and syrup and vanilla cream, but I wouldn't turn down a good bacon, scrambled egg and syrup waffle, either. There's a bit at the end of Nora Ephron's final book where she's describing things she'll miss when she dies, and she lists both waffles and the concept of waffles. I couldn't agree more." – **Ruby Tandoh, food writer, Sheffield, UK**

"Definitely fresh coffee and fresh-squeezed orange juice. If it's winter, a bowl of Macroom oatmeal with jersey cream and soft brown sugar, then some Ballymaloe sourdough bread with jam or marmalade. I love in particular homemade kumquat marmalade, or real, dark, bitter Seville orange marmalade. On weekends it's nice to have a rasher or two with everything else – the full works." – **Darina Allen, food writer & educator, Cork, Ireland**

THE PALM TEST FOR STEAK

It's a neat idea: you can tell how well-done a steak is by comparing its firmness to the palm of your hand held in various configurations. The steak is (a) rare, (b) medium or (c) well-done if it feels like the thumb-edge of your palm when you touch your thumb to your (a) index finger, (b) middle finger or (c) ring finger. Got it?

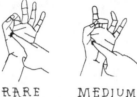

RARE   MEDIUM   WELL

However, as J Kenji Lopez-Alt of *Serious Eats* illustrates, the test is practically useless. "First off, not all hands are created equal … Then we get to the meat itself. Thick steaks don't compress the same way as thin steaks. Fatty steaks don't compress the same way as lean steaks. Tenderloins don't compress like ribeyes." The only reliable method of testing the doneness of a steak is to use an accurate, instant-read kitchen thermometer (Kenji recommends the Thermopen).

SPECIALIST FOOD FESTIVALS

Festival & Country	Speciality
Yulin festival, *China*	Lychees and dog meat
Hokitika Wildfoods Festival, *New Zealand*	Rarely eaten foods
Blue Food Festival, *Tobago*	Blue food[1]
Watercress Festival, *England*	Watercress
La Tomatina, *Spain*	Tomatoes (thrown)
Annual Golden Spurtle, *Scotland*	Porridge
PoutineFest, *Canada*	Poutine
Pizzafest, *Italy*	Pizza
Zibelemärit, *Switzerland*	Onions
Battle of the Oranges, *Italy*	Oranges (thrown)
Castagnades d'Automne, *France*	Chestnuts

[1] The local name for a root vegetable called *dasheen*, which turns a blueish colour when boiled.

NASA'S CORNBREAD

For consumption in outer space (*see* page 76) with Thanksgiving turkey.

Ingredient	%	Amount (g)
Chicken broth	40.90	899.80
Cornbread, prepared, crumbled	39.50	869.00
Onions, chopped	10.47	230.34
Celery, chopped	6.15	135.30
Butter, unsalted	2.52	55.44
Salt	0.20	4.40
Poultry seasoning	0.11	2.42
Black pepper	0.07	1.54
Parsley flakes, dried	0.04	0.88
Sage, rubbed	0.04	0.88
Total	**100.00**	**2200.00**

Directions (earth kitchen): Preheat convection oven to 160°C (325°F/gas mark 3). Conventional oven should be heated to 180°C (350°F/gas mark 4). Grease 22 × 33-cm (9 × 13-inch) baking pan. Peel onions and purée in food processor. Place in bowl. Set aside. Finely chop celery in food processor. Add to onion purée. Set aside. Heat sauté pan over medium heat. Melt butter and sauté onion and celery mixture until soft (about 5 minutes). Add to crumbled, prepared cornbread. Mix well. In a separate bowl, combine salt, poultry seasoning, black pepper, parsley and sage. Add to cornbread-sautéed vegetable mixture. Add chicken broth. Mix well. Spoon dressing into prepared baking pan. Bake for approximately 35 minutes.

For space flight preparation: Baked dressing is transferred to metal tray and freeze-dried accordingly. One serving of cornbread dressing shall weigh approximately 145g (5¼oz) prior to freeze-drying and 50g (1¾oz) after freeze-drying.

Please note: This recipe is based on the "formulations" for actual space flight missions. Actual measures are estimates. It should also be noted, "space flight food" recipes are designed with significant flavouring to compensate for the freeze-drying process.

DESERT ISLAND DISHES

Since 1942, the BBC Radio 4 programme *Desert Island Discs* has been dispatching guests, or "castaways", to an imaginary desert island with just a few personal effects to tide them over: eight recordings, the *Complete Works of Shakespeare*, the *Bible* (or other religious work), a third book of choice and, since 1951, a luxury that is (a) inanimate and (b) won't facilitate communication with the outside world. Of the 3,000-plus guests who've appeared on the show, more than 400 have chosen food- or drink-related items as their luxury – some of them more practical than others.

Year	Castaway	Luxury
1951	Sally Ann Howes	Garlic
1952	Delia Murphy	Still for making poteen
1952	Trevor Howard	One of his wife's cakes
1956	Shirley Abicair	Case of avocado pears
1957	Commander Ibbett	Barrel of beer
1958	Jean Sablon	Menu from Maxim's in Paris
1959	George Melachrino	Turkish delight
1960	Alec Guinness	Apricot brandy
1960	Johnny Morris	Yeast
1964	Vanessa Redgrave	Coffee and condensed milk
1965	Basil Spence	Spaghetti
1965	Anatole de Grunwald	Caviar
1965	Sheila Hancock	Cat food
1967	Renée Houston	Parsley
1971	Ludovic Kennedy	Tartar sauce
1975	Alan Civil	Something to improve the taste of coconut milk
1976	Sherrill Milnes	Herrings in sour cream
1977	Jessica Mitford	Supply of Gentleman's Relish
1986	Dennis Taylor	Limitless supply of yogurt
1988	Germaine Greer	Hot spices
1991	Sue Townsend	Swimming pool of Champagne
1994	Ian Hislop	Frosties cereal
1994	Howard Hodgkin	Mayonnaise – permanent supply
1999	Fay Maschler	Huge supply of ouzo
1999	Rick Stein	Thai fish sauce

2000	Norman Wisdom	Pot of stew with two dumplings
2001	Joss Ackland	Huge jar of liquorice
2002	Gordon Ramsay	Fresh vanilla pod
2003	Vic Reeves	Potato seeds
2004	Ann Leslie	Enormous amount of garlic
2005	Carlos Acosta	Case of Havana rum
2005	Boris Johnson	Large pot of French mustard
2006	Frankie Dettori	Lifetime's supply of Pinot Grigio
2006	Terence Stamp	One of his wheat-free loaves
2006	Armando Iannucci	Virtual sherry trifle
2006	Heston Blumenthal	Japanese knives
2007	Wangari Maathai	Huge basket of fruit
2007	Thomas Keneally	Can of caviar, spoon, tin opener
2007	Andrew Davies	Endless supply of mojitos
2008	Howard Goodall	Ice-cold vanilla vodka and tonics
2008	Peter Carey	"Magic" pudding and a drink
2008	Antonio Carluccio	White truffles
2009	Whoopi Goldberg	Wise potato chips
2009	Jerry Springer	Cheeseburger machine
2012	Tidjane Thiam	Solar-powered ice-cream maker
2013	Ed Miliband	Weekly delivery of chicken tikka
2015	Ruth Rogers	Felsina or Fontodi olive oil
2016	Jilly Cooper	Sack of nuts
2016	Yotam Ottolenghi	Lemon tree
2017	Arundhati Roy	Ratol mango tree

MOST-REQUESTED DESERT ISLAND DISCS LUXURIES (FOOD & DRINK)

Wine (75)

Champagne (46)

Whisky (33)

Tea (30)

Coffee (23)

Seeds to grow food (19)

Alcohol still (11)

Chocolate (10)

Fishing gear (10)

Brandy (9)

Beer (8)

Ice cream (5)

Marmite/Vegemite (5)

Caviar (4)

Olive oil (4)

ON EATING ALONE

Solo dining is on the rise. As Western society atomizes – 60 per cent of Stockholm residents now live alone – the stigma once attached to solitary eaters is fading. The emphasis on bar seating in many new restaurants encourages the practice and there is an Amsterdam restaurant, Eenmaal, which only caters for parties of one[1]. Whether or not eating alone is a positive thing has long been up for debate.

IN FAVOUR

"There are few people alive with whom I care to pray, sleep, dance, sing, and (perhaps most of all, except sleep) share my bread and wine. Of course there are moments when such unholy performances must take place, in order to exist socially, but they are endurable because they need not be the only fashion of self-nourishment."

MFK Fisher
An Alphabet for Gourmets

AGAINST

"It is the saddest sight in the world. Sadder than destitution, sadder than the beggar is the man who eats alone in public. Nothing more contradicts the laws of man or beast, for animals always do each other the honour of sharing or disputing each other's food. He who eats alone is dead (but not he who drinks alone. Why is this?)."

Jean Baudrillard
America

[1] On the other side of the debate, the Moomin House Café chain in Japan takes pity on solo diners, seating them with giant stuffed animals as a way of warding off loneliness.

ESCANCIAR

Walk into a *sidreria*, or cider bar, in Spain's northwestern province of Asturias and you'll encounter the very curious sight of a waiter lifting a bottle high in the air, tilting it and letting a long stream of natural cider fall into a glass below. This method, which aerates the still liquid, bringing it to life for a few precious moments, is known as *escanciar*.

THE GANNET EXPLAINS... MSG

What is it?

A key component in Chinese cooking, monosodium glutamate (MSG) was first prepared in the early 1900s by Japanese biochemist Kikunae Ikeda, who was studying the flavour of dashi broths. He isolated glutamic acid, which occurs naturally in seaweed, tomatoes, mushrooms and other foods, and stabilized it with salt. The resulting white powder quickly became popular in many Asian countries.

Because it's delicious?

Yes, and moreish too. MSG is rich in umami, a Japanese term usually equated with "savouriness". Chefs use it to deepen the flavour of a dish and provide a savoury kick.

So what's the problem?

It's commonly believed to cause headaches and other feelings of discomfort (weakness, numbness, heart palpitations). What's come to be known as "Chinese-restaurant syndrome" began with a letter from Dr Robert Ho Man Kwok, a Chinese-American biomedical researcher, to the *New England Journal of Medicine* in 1968. Feeling unwell after a Chinese meal, he speculated that MSG might be the cause. (He also suspected Chinese cooking wine and the high sodium content in some dishes.) The media picked up the story and ran with it, leading to decades-long suspicion of the salt – many restaurants are still afraid to use MSG for this reason.

So we should avoid MSG at all costs?

Not so, according to top chefs David Chang and Heston Blumenthal, and star food scientist Harold McGee, who argue that the stigma attached to MSG has no basis in fact (some even suspect it has racist undertones). "MSG is a common ingredient in foods, and not just in Chinese restaurants; it occurs naturally in many of the foods we love," writes McGee, noting that after hundreds of scientific studies around the world, there is no evidence that MSG causes the symptoms of Chinese Restaurant Syndrome.

THE WORLD'S FIRST TEA BOOK

Ch'a Ching, or *The Classic of Tea*, by Lu Yu, a writer raised in a Buddhist monastery in 8th-century China, is the earliest known monograph on tea. It provides an extremely comprehensive guide to all aspects of tea-making, from soil to harvesting, to the final pour. To brew and serve tea properly, according to Lu Yu, one needs the following pieces of equipment.

Brazier (brass or iron)	Water filter (raw copper)
Basket (bamboo)	Water ladle (split gourd)
Stoker (iron)	Bamboo pincers (peach wood)
Fire tongs (iron/copper)	Salt dish (stoneware)
Cauldron (preferably silver)	Heating basin (stoneware/clay)
Stand (in the shape of a cross)	Tea bowl (Yueh chou ware)
Pincers (green bamboo)	Basket for cups (white rush)
Paper sack (rattan)	Brush (coir palm bark)
Roller & brush (wood, feathers)	Scouring box (catalpa wood)
Gauze & casket (silk, bamboo)	Container for dregs (wood)
Measure (sea shell)	Cloth (coarse thread)
Water dispenser (catalpa wood)	Utensil rack (wood/bamboo)
	Carryall (bamboo)

Lu Yu also had a lot to say about the proper type of water to use for brewing tea. "I would suggest that tea made from mountain streams is best, river water is all right, but well-water tea is quite inferior," he wrote. "Water from the slow-flowing streams, the stone-lined pools or milk-pure springs is the best of mountain water. Never take tea made from water that falls in cascades, gushes from springs, rushes in a torrent or that eddies and surges as if nature were rinsing its mouth. Over usage of all such water to make tea will lead to illnesses of the throat… If you must use river water, take only that which man has not been near; and if it is well water, then draw a great deal before using it."

The significance of tea in Chinese life can be felt in this poem about Lu Yu by one of his contemporaries:

The Day I Saw Lu Yu off to Pick Tea

A thousand mountains will greet my departing friend,
When the spring teas blossom again.
With such breadth and wisdom,
Serenely picking tea—
Through morning mists
Or crimson evening clouds—
His solitary journey is my envy.
We rendezvous at a remote mountain temple,
Where we enjoy tea by a clear pebble fountain.
In that silent night,
Lit only by candlelight,
I struck a marble bell—
Its chime carrying me
A hidden man
Deep into thoughts of ages past.
~ Huangfu Zheng

THE VATICAN WINE MYSTERY

According to the California-based Wine Institute, the Vatican City drinks more wine per capita than any other country on earth: on average each citizen of the tiny state drank 74 litres (156 pints) – or 98 regular-sized bottles – in 2012. Why might this be?

Theory #1: The average is inflated by communion wine.
Rebuff: "That can't be right," columnist Michael Winterbottom of *The Universe* Catholic weekly told the *Guardian*. "Most of the time, you don't have wine with communion."

Theory #2: According to *La Stampa*, there is a supermarket in the Vatican where low tax rates are exploited by pass holders.
Rebuff: The supermarket is tiny, as Winterbottom points out, and the wine there is no cheaper than elsewhere.

Theory #3: The Vatican is child-free. No other country on earth is entirely populated by adults of legal drinking age. QED.

ARTISTS IN THE KITCHEN

It's one thing to paint a bowl of fruit or screenprint a soup can, and quite another to roll up your sleeves and actually cook an elaborate feast (or at least provide the instructions for someone else to do so) as part of your artistic practice. Here are some prime examples from the past 100 years of artists taking on chefs at their own game.

» The founder of the Futurist movement **Filippo Marinetti** wasn't just out to revolutionize the arts and lead the world into a glorious new mechanical age of fast cars and air travel. He also had big ideas for the future of food – not least his plan to abolish pasta, aka "the dictator of the stomach", which didn't go down so well when he announced it in his native Italy. To disseminate his culinary ideas, Marinetti staged a series of very unusual banquets around Europe. At the Penne d'Oca restaurant in Milan in 1930, diners were bathed in green light, poured wine from jerrycans labelled "Extra-thick oil" and sprayed with strong carnation perfume. Subsequent events showcased dishes such as "Raw Meat Torn by Trumpet Blasts" – diners were instructed to blow a trumpet between mouthfuls of electrified beef – and "The Excited Pig", for which a whole salami was cooked in strong espresso and flavoured with eau-de-cologne. (These and more recipes are gathered in *The Futurist Cookbook* of 1932.) It would be easy to dismiss all this as one big zany provocation, but it's possible to see traces of Marinetti's culinary legacy in the highly technical, often shocking dishes of present-day chefs such as Ferran Adrià and Heston Blumenthal (*see* page 7).

» Another artist who blurred the boundaries between food and art was **Salvador Dalí**, whose extraordinary series of dinners in Paris in the early 1970s was surely influenced by Marinetti's banquets four decades earlier. Held in secret at top Paris restaurants including Maxim's, these feasts were later revealed in a lavish hardback called *Les Dîners de Gala* (1973, republished 2016). Here you'll find photographs of Dalí posing next to wildly extravagant spreads and recipes like Thousand-Year-Old Eggs, Peacock à l'Impériale (actually made with quail) and Conger of the Rising Sun.

» The New York restaurant Food, which popped up in SoHo in 1971 and ran for three years, was nothing if not ahead of its time. Founded by artists **Carol Goodden**, **Tina Girouard** and **Gordon Matta-Clark**, it made a point of serving fresh and seasonal food, a commonplace decision now but quite radical then. They conceived of cooking as a kind of performance: the kitchen was open to the dining room and artists were invited to appear as guest chefs. For one bone-themed dinner, priced at $4, Matta-Clark served oxtail soup, roasted marrow bones and frogs' legs, then had the used bones cleaned and turned into necklaces for his guests to wear.

» In the 1990s, the Buenos Aires-born artist **Rirkrit Tiravanija** turned a series of New York galleries into kitchens. The work was the cooking itself – *pad Thai* for one show in 1990, Thai curry and rice for another – and gallery visitors literally consumed the art. Tiravanija wanted people to interact with contemporary art in a more sociable way, blurring the distance between artist and viewer.

» New York artist **Jennifer Rubell** worked in food – as a hotelier, then a food writer – before she started working *with* it as an artist, staging bizarre buffets at art events and edible installations in galleries. Her opening-night dinner at Performa 2009 in New York involved one ton of barbecued ribs, three felled apple trees and seven chocolate mini-facsimiles of Jeff Koons's rabbit sculpture.

CHEFS MAKING ART

The list of chefs who have proved themselves as artists is considerably shorter than the converse, though there is a case to be made that cooking at a certain level could be perceived as art. This argument was advanced in 2014 by **Ferran Adrià** who, following the closure of elBulli, toured an exhibition of his work – sketches, charts, shopping lists and a video of nearly 2,000 dishes – around galleries in Europe and the US. Elsewhere, chefs such as **David Muñoz** at DiverXO in Madrid and **Grant Achatz** at Alinea in Chicago use tables like canvases on which to "paint" Pollock-style dishes for their customers – look up "Alinea dessert" on YouTube and decide for yourself whether it qualifies as art.

BE CAREFUL WHAT YOU ORDER

Not all dish names are literal and sometimes the thing arriving on your table is alarmingly different from what the wording led you to expect. Often this is a result of linguistic or historical distance – the name sounded perfectly reasonable at its point of origin – but in other cases the innuendo is completely intentional.

EUPHEMISTIC

Rocky Mountain oysters – A nice American way of saying you're eating bull, pig or sheep testicles – peeled, flour-coated, deep-fried and served with cocktail sauce.

Bombay duck – Poultry-lovers will be perplexed to discover that this is, in fact, a rather unsightly type of lizardfish. Popular in Mumbai, it is usually dried and salted before consumption and gives off a pungent fishy reek.

Sweetbreads – Not something to nibble with your morning coffee, but rather the inner organs and glands of calf and lamb (and sometimes cow and pig), often served deep-fried.

DYSPHEMISTIC

Spotted dick – A perfectly presentable British suet pudding. The dried fruit (currants, raisins) account for the "spotted" part, though "dick" is harder to guess

– it may be an old Yorkshire term for pudding.

Festy cock – Also known as a "fitless cock", this is a kind of Scottish pancake made with oatmeal, moulded into the shape of a cockerel and baked.

Stinking Bishop – It's got a strong aroma, sure, but this Gloucestershire cows' cheese has a very fine flavour. The name comes from a local pear bred by an allegedly angry 19th-century farmer named Frederick Bishop.

RELIGIOUS

Priest stranglers – The name of a twisty pasta popular in Emilia-Romagna, *strozzapreti* may allude to a possible result of clerical overindulgence.

The Imam fainted – The origin of *Imambayıldı* is obscure but this Turkish dish of split aubergine with tomato, garlic and onion is swoonsome.

THINGS THAT SHOULDN'T BE KEPT IN THE FRIDGE

Food	Reason
Tomatoes	Enzymes break down, killing the flavour
Bread	Dries out and goes stale faster
Bananas	Turn mushy and black
Potatoes	Starch turns to sugar, which produces a potentially harmful chemical when baked or fried
Eggs	May thin the whites, other flavours may penetrate[1]
Honey	Crystallizes, forming a grainy texture
Coffee	May take on odours from other foods
Onions	May taint the flavour of other foods
Garlic	May go mouldy
Basil	Wilts faster, absorbs other odours
Open cans	Metals from the can may leach into the contents
Hot food	Causes condensation, raises fridge temperature
Whole melons	Lose antioxidants, go off quicker
Avocados	Ripening is prevented

[1] There is an ongoing debate around egg refrigeration. Many health bodies – particularly in the US, where hens are not routinely immunized against salmonella as they are in Europe – caution that keeping eggs at room temperature raises the risk of food poisoning. In the UK, some authorities discourage refrigeration on the basis that chilling and then warming eggs could create condensation, which would allow salmonella to penetrate the shell.

FUNISTRADA

In 1974, the US army carried out a survey of soldiers' dietary preferences. When the results were published, something called funistrada ranked higher than fried aubergine (eggplant) and onion soup. Funistrada does not in fact exist: it was included in the survey, along with buttered ermal and braised trake, "to provide an estimate of how much someone will respond to a word which sounds like a food name or will answer without reading".

HIDDEN DANGERS

"Everything is poison and nothing is without poison, it is only the dose that makes some things not a poison," wrote the Swiss-German philosopher and toxicologist Paracelsus (1493–1541). In other words, don't get distracted by the flamboyantly poisonous things in nature, for even the seemingly benign foods we eat on a daily basis may secretly have it in for us. The answer in most cases, Paracelsus would agree, is to take everything in moderation.

Bananas – You would have to eat a stupendous number of bananas to die of radiation poisoning – one Harvard doctor puts it at ten million – but it's possible in theory as every banana contains a tiny amount of radioactive potassium (about one microsievert). Bananas are sometimes used to measure radiation. Two weeks after the Japanese nuclear power plant at Fukushima was hit by a tsunami, the local town hall received a BED (Banana Equivalent Dose) of 1,000 – the same exposure a human would get from eating 1,000 bananas.

Ackee – To safely eat the national fruit of Jamaica (usually paired with saltfish), you have to skirt a couple of pitfalls. The black seed within the pod is poisonous. So is the edible part – the yellow arilli – if the fruit isn't fully ripe. The poison hypoglycin causes Jamaican vomiting sickness, which can be fatal. In 2000–1, over 100 people reportedly died in Haiti in just four months from ackee poisoning.

Apricots – Apples usually get a bad rap for containing arsenic in their pips – or to be exact, a molecule called amygdalin, which can produce hydrogen cyanide through digestion – but apricot seeds contain nearly five times as much amygdalin. In either case, you'd have to eat a *lot* of seeds to poison yourself.

Cassava – The third-largest source of edible carbohydrate in the world, this woody shrub (also known as yuca or manioc) is nevertheless poisonous if eaten raw or inadequately cooked. It contains several types of cyanide in the roots, peel and leaves, and the bitter cultivar can yield up to 1 gram of cyanide per kilo (2¼ lb) – just a few pieces of root can deliver a fatal dose. Uncooked cassava is also linked to a disease called konzo, which causes paralysis.

TASTE vs FLAVOUR

Taste refers to the five qualities detectable in the mouth: sweetness, sourness, saltiness, bitterness and umami.

Flavour is the overall sensory impression of food or other substances determined primarily by our senses of taste and smell.

Our experience of flavour can also be affected by temperature, texture and the trigeminal senses, which detect chemical irritants in the mouth and throat (heat from chillies, the cooling properties of menthol, the puckering effect of tannins in red wine).

SOME EVOCATIVELY NAMED CHILLI SAUCES

When your chilli-based product is ridiculously hot, it helps to broadcast this by giving it a ridiculous name. Here are a few of the more eye-catching examples (many of which contain "superhot" chillies as detailed on page 72).

Crazy Bastard Sauce

Dave's "Insanity" Sauce

Endorphin Rush "Beyond Hot" Sauce

Grinders "Near Death" Hot Sauce

Xtinction Sauce with Ghost Pepper

Marie Sharp's "Beware" (Comatose Heat Level)

Dragon's Blood Hot Naga Sauce

Chilli Pepper Pete's "Hotter Than Hell"

"The Widowmaker" California Reaper Hot Sauce

Lethal Ingestion Hot Sauce

Blair's "Ultra Death" Sauce

Crazy Jerry's "Brain Damage" Hot Sauce

Ass Blaster Hot Sauce (comes in wooden "outhouse")

Doyle Wolfgang von Frankenstein's "Made in Hell" Hot Sauce

ABSURDIST MOMENTS IN FOOD SCIENCE

A parody of the Nobel prizes, the Ig Nobels were established in 1991 "to celebrate the unusual, honour the imaginative, and spur people's interest in science, medicine, and technology," according to co-sponsor Marc Abrahams. Their motto: "First make people laugh, and then make them think." A few of the awards have implications for food and drink...

2015

Callum Ormonde and Colin Raston, and Tom Yuan, Stephan Kudlacek, Sameeran Kunche, Joshua N. Smith, William A. Brown, Kaitlin Pugliese, Tivoli Olsen, Mariam Iftikhar and Gregory Weiss won the Chemistry prize **for inventing a chemical recipe to partially un-boil an egg.**

2014

Jiangang Liu, Jun Li, Lu Feng, Ling Li, Shubham Bose, Jie Tian and Kang Lee won the Neuroscience prize **for trying to understand what happens in the brains of people who see the face of Jesus in a piece of toast.**

2013

Brian Crandall and Peter Stahl won the Archaeology prize **for parboiling a dead shrew, swallowing it without chewing, and then carefully examining everything excreted during subsequent days, so they could see which bones would and would not dissolve inside the human digestive system.**

2009

Stephan Bolliger, Steffen Ross, Lars Oesterhelweg, Michael Thali and Beat Kneubuehl of the University of Bern, Switzerland, won the Peace prize **for determining whether it is better to be hit on the head with a full bottle of beer or with an empty bottle.**

2008

Massimiliano Zampini of the University of Trento, Italy, and Charles Spence of Oxford University, UK, won the Nutrition prize **for electronically modifying the sound of a potato chip to make the person chewing the chip believe it to be crisper and fresher than it really is.**

2008

Sharee A. Umpierre of the University of Puerto Rico, Joseph A. Hill of The Fertility Centers of New England (USA), Deborah J. Anderson of Boston University School of Medicine and Harvard Medical School (USA) won the Chemistry prize **for discovering that Coca-Cola is an effective spermicide** – shared with Chuang-Ye Hong of Taipei Medical University (Taiwan), C.C. Shieh, P. Wu and B.N. Chiang (all of Taiwan) **for discovering that Coca-Cola is not an effective spermicide.**

2006

Antonio Mulet, José Javier Benedito and José Bon of the University of Valencia, Spain, and Carmen Rosselló of the University of Illes Balears, in Palma de Mallorca, Spain, won the Chemistry prize **for their study *Ultrasonic Velocity in Cheddar Cheese as Affected by Temperature.***

2004

Jillian Clarke of the Chicago High School for Agricultural Sciences, and then Howard University, won the Public Health prize **for investigating the scientific validity of the Five-Second Rule about whether it's safe to eat food that's been dropped on the floor** (*see* page 9 for more on this).

GARUM

The flavour of fish sauce, a condiment extracted from brine-fermented fish, is something we associate with Southeast Asian cuisine, but in fact it was a key part of cooking in Ancient Greece and Rome. Whole small fish such as sand smelt or anchovies, or the entrails of larger fish like tuna, were used to make garum, prized for its salty, savoury taste. The tradition vanished in medieval times, but is slowly creeping back into fashion: a few drops of *colatura di alici*, which you can find online, goes amazingly well with tomatoes or steak.

HITLER'S FOOD TASTER

In winter 1941, a young woman named **Margot Wölk** fled her bombed-out apartment in Berlin and sought refuge at her mother-in-law's home in East Prussia. It so happened that, just three kilometres (nearly two miles) away, Adolf Hitler had moved into the Wolf's Lair, his secret military headquarters on the Eastern Front, where he would remain until November 1944. Hitler was acutely paranoid about being poisoned, and so it was arranged that 15 local women, including Wölk, would serve as his food tasters. Each morning, they were bussed to a nearby school, where platters of vegetables – "the most delicious fresh things, from asparagus to peppers and peas, served with rice and salads" – were laid out. Only after the women had tasted food from every plate would it be relayed on to the Wolf's Lair.

Wölk was not a Nazi and, like the other women, was forced to do the job against her will. "We had to eat it all up," she said in an interview in 2014. "Then we had to wait an hour, and every time we were frightened that we were going to be ill. We used to cry like dogs because we were so glad to have survived." This went on without incident until late 1944, when Wölk managed to return to Berlin. She never glimpsed Hitler during those three years, and of the 15 food tasters she was the only one to survive the war – the rest were reportedly shot by the Red Army in January 1945.

OTHER WORLD LEADERS WHO EMPLOYED FOOD TASTERS

Emperor Claudius – Died from mushroom poisoning in 54BC. His taster, Halotus, was a suspect though he survived to become a taster for Claudius's successor, Nero.

Nicolae Ceaușescu – The Romanian dictator brought a food taster with him on a state visit to Buckingham Palace in 1978.

Saddam Hussein – Had several tasters. One of them, Kamel Hana Gegeo, was bludgeoned to death by Saddam's son Uday in 1988.

Vladimir Putin – Employs a full-time food taster and has his meals prepared by a member of his security staff.

UNDERCOVER FRUITS AND BOGUS BERRIES

Everyone knows that the tomato, despite being treated as a vegetable in most kitchens, is actually a fruit[1]. From a botanical perspective, the following also qualify:

Bell pepper	Okra	Olive
Cucumber	Sunflower	Pumpkin
String beans	seeds	Avocado
Corn	Squash	Nuts

The confusion is due to the fact that while "fruit" is a botanical term, "vegetable" is not – it's a food group defined by cultural and culinary tradition – so it's possible to call something a fruit and a vegetable at the same time without contradicting yourself.

A similar confusion arises around the classification of berries, which are defined as a fleshy fruit formed from the ovary of a single flower with seeds embedded *inside* the flesh. Rather upsettingly, this excludes strawberries, raspberries and blackberries, all of which have their seeds on the outside. You are to think of these as "aggregate fruits" – they develop from multiple ovaries of a single flower. To confuse matters further, a number of very un-berry-like fruits are technically berries:

Banana	Watermelon
Avocado	Kiwi
Tomato	Pumpkin

You may never look at your strawberry jam in the same way again.

---

[1] It develops from the ovary in the base of the flower and contains the seeds of the tomato plant. The US Supreme Court doesn't recognize its fruitiness, however. In 1893, a fruit importer brought a case against the Port of New York challenging the classification of tomatoes as vegetables, which were subject to import tariffs. The court rejected the case and continues to classify tomatoes as they are defined "in the common language of the people" given that they are "usually served at dinner in, with, or after the soup, fish, or meats which constitute the principal part of the repast, and not, like fruits generally, as dessert".

THE BUILDING BLOCKS OF GIN

Before visiting the **Bertha's Revenge** distillery in Co Cork, Ireland – established in 2015 by hotelier Justin Green and wine trader Antony Jackson – I had no idea how many ingredients could go into a bottle of gin. Bertha's Revenge contains the following:

Whey alcohol	Grapefruit	Orris	Cumin
Spring water	Sweet orange	Angelica	Almond
Juniper	Lemon	Cinnamon	Elderflower
Coriander	Lime	Cardamom	Alexanders
Bitter orange	Liquorice	Cloves	Sweet woodruff

The various seeds, berries, herbs, roots and fruits used for flavouring are known as botanicals, juniper being the most important. Some gin-makers use botanicals with restraint; some (particularly new-breed distillers like Bertha's Revenge) are more liberal. Other acceptable botanicals include: anise, cassia bark, honeysuckle, frankincense, olives, kaffir lime leaves, yarrow, hops, cubeb berries, hibiscus and grains of paradise.

BERTHA'S REVENGE RHUBARB MARTINI

Serves 2

75ml (2½fl oz) gin
50ml (2fl oz) rhubarb syrup
(see below)
12.5ml (2½ teaspoons)
lemon juice

Rhubarb syrup

75g (2½oz) sugar
75ml (2½fl oz) water
300g (10½oz) rhubarb,
trimmed and chopped

To make the syrup, gently heat the sugar and water in a pan over a low heat until the sugar dissolves. Add the rhubarb, bring to the boil and simmer, covered, until the rhubarb is very tender (about 5 minutes). Remove from the heat and pass through a sieve, then return the juice to the pan and boil for a few minutes until just syrupy. Pour into a jug – you should have about 125ml (4fl oz) – and leave to cool.

Fill a cocktail shaker nearly to the top with ice. Add the gin, rhubarb syrup and lemon juice and shake vigorously for 30 seconds. Strain the drink into 2 chilled glasses. Taste, adjust if necessary and serve.

...AND OTHER ALCOHOLIC DRINKS

Not all libations are as multifarious as gin, and some can be boiled right down to just a couple of ingredients. It should be noted that this list covers only the minimum requirements; many of the following can be complicated in numerous ways.

DISTILLED

Whisky
Grains (barley, corn, rye, wheat) + yeast + water

Vodka
Grains (rye, barley, wheat) or potatoes + yeast + water

Rum
Molasses or sugarcane juice + yeast + water

Brandy
Grapes + yeast

Tequila
Blue agave juice + yeast + water

Slivovitz
Damson plums + yeast + sugar

Shochu
Grains + koji[1] + water

Absinthe
Herbs (incl. anise, fennel) + botanicals (incl. wormwood) + alcohol

FERMENTED

Wine
Grapes + yeast

Beer: lager
Grains + water + hops + bottom-fermenting yeast

Beer: ale
Grains + water + hops + top-fermenting yeast

Cider
Apples + yeast

Perry
Pears + yeast

Sake
Rice + koji + yeast + water

Port
Grapes + yeast + grape spirit or brandy

Tonto (banana beer)
Bananas + sorghum

Mead
Honey + yeast + water (+ grains)

---

[1] Koji is a mould used for fermenting and saccharifying in various East Asian cuisines.

GASTRONOMIC KILLERS

We need it to live, but the food we eat has always been shadowed by death. Certain things are obviously dangerous but people eat them, out of ignorance, desperation or, in the case of *fugu*, as a gastronomic Russian roulette. But benign foods can prove unexpectedly fatal. Here are a number of famous deaths and the food or drink that caused them.

Mushrooms – A veritable serial killer of the food world, fungi have been held responsible for the deaths of the Roman Emperor Claudius, Pope Clement VII, Holy Roman Emperor Charles VI and Tsaritsa Natalya Naryshkina (all from eating death cap mushrooms). Some say Siddhartha Gautama (aka the Buddha) also died from mushroom poisoning.

Lampreys – According to a contemporary chronicle, King Henry I of England, youngest son of William the Conqueror, died of a "surfeit of lampreys" while on a hunting trip in Normandy – though most historians believe he died of food poisoning.

Chicken – Chicken can be dangerous if insufficiently cooked, but this did not cause the death of the English philosopher Francis Bacon in 1626. Rather, he died from a chill caught while stuffing a chicken with snow, during a very forward-thinking experiment with refrigeration.

Fugu – In 1975, the Japanese kabuki actor Bando Mitsugoro VIII expired from eating four servings of blowfish liver. Despite containing tetrodotoxin, a poison 1,000 times deadlier than cyanide, *fugu* is still popular in Japan – 10,000 tons of the fish are consumed each year.

Cherries and milk – In 1850, US President Zachary Taylor consumed a large quantity of cherries and iced milk at a Fourth of July celebration in Washington, fell sick shortly after, and died on 9 July. His physicians attributed his death to cholera, but others claim it was gastroenteritis caused by highly acidic cherries mingling with fresh milk.

Poisoned apple – Biographers of Alan Turing have claimed that the mathematician killed himself with a poisoned apple, inspired by a scene from his favourite fairy tale *Snow White and the Seven Dwarfs*, but this has been disputed. For one, the half-eaten apple found by his bedside was never tested for cyanide. A counter-theory suggests he poisoned himself by accident while working on an experiment, perhaps by inhaling cyanide fumes.

Flavor Aid – Although the Kool-Aid brand is popularly associated with the 1978 mass murder and suicide in Jonestown, Guyana, it was in fact Flavor Aid (the grape variety) laced with arsenic which killed 918 members of the Peoples Temple, an American religious organization. Neither drink killed its leader Jim Jones: he died from a shot to the head, probably self-inflicted.

White mullet – This seemingly benign fish was blamed for the death of Aimery, the first king of Cyprus, in 1205 – the reported cause was overeating.

Malmsey wine – Alcohol has been the sad cause of many famous deaths throughout history, but relatively few have died by drowning in it, as George Plantaganet, Duke of Clarence, is believed to have done in 1478. Legend has it that he was executed by immersion in a butt of Malmsey wine during his imprisonment in the Tower of London.

Semla – King Adolf Frederick is remembered by Swedish school children as the king who ate himself to death. He died in 1771 after a meal which included lobster, caviar, sauerkraut, kippers and Champagne, followed by 14 servings of his favourite dessert, *semla* (a sweet roll), served in a bowl of hot milk.

HAENYEO

For centuries, the sea women, or *haenyeo*, of South Korea have been diving for seafood in the ocean off Jeju province. They descend to depths of 10m (33ft) without breathing equipment, armed only with scuba masks, lead weights and a tool for digging abalone and other edible creatures off the sea floor. A typical dive lasts about two minutes and they tend to work for five or six hours at a stretch, though time may be running out for the profession as a whole: approximately 2,500 *haenyeo* remain today, down from more than 20,000 in the 1960s, and the vast majority of them are now over the age of 60.

HOLLYWOOD LOVES
JAPANESE WHISKY

Two decades before Sofia Coppola made *Lost in Translation* (2003), in which an ageing American movie star played by Bill Murray arrives in Tokyo to film a commercial for Suntory whisky, Sofia's father, Francis Ford Coppola, and the great Japanese filmmaker, Akira Kurosawa, filmed a series of commercials for the very same whisky brand. ("There's no stronger friendship than that between these two men," the narrator intones, as the directors sip Suntory Reserve.) Far from being a one-off, these ads are part of a long and distinguished relationship between Hollywood stars and Japanese whisky brands.

Orson Welles – The legendary actor/director flogged everything from frozen peas to photocopiers. In 1979, it was the turn of Nikka's G&G whisky. "What we're always trying for of course is perfection," Welles tells the audience with a derisive grunt. "In a film that's only a hope, but with G&G you can rely on it." He doesn't look entirely convinced.
Dignity: 3/10

Sammy Davis Jr. – Pouring himself a glass of whisky, the entertainer plays the bottle like a bass guitar and improvises a groovy scat, before delivering the only word in this 1974 advert: "Suntory". No translation required.
Dignity: 9/10

Mickey Rourke – No words at all were required from the heart-throb actor when he filmed a couple of commercials for Suntory Reserve in the 1980s. He merely had to don a dinner jacket, sip some whisky and look vaguely interested.
Dignity: 5/10

Sean Connery – Perhaps the least artistically credible examples of the genre, Connery's 1992 ads for Suntory Crest are full of plaintive piano, purple shirts and soft-focus shots of beaches – although the Bond actor does raise one very sardonic eyebrow during a reaction shot, signalling that he's well aware of the absurdity of it all.
Dignity: 2/10

Roy Scheider – The *Jaws* star chews an ice cube and pulls weird faces at a golden retriever in a very offbeat early-1980s ad for Super Nikka. The payoff: "I'm a boy at heart."
Dignity: 4/10

Keanu Reeves – Lightning flashes, Keanu making music in a darkened mansion, a cat that morphs into a predatory brunette. This moody 1990s ad for Suntory Reserve ramps up the sexiness but resolutely fails to keep silliness at bay. Key shot: the soon-to-be *Matrix* star falling backwards over a couch as he retreats from the prowling cat/woman.
Dignity: 2/10

MORE WEIRD FOOD & DRINK ENDORSEMENTS

Paul Newman
Product: Maxwell Blendy coffee (Japan)
Plot: Woman drops cake. Newman, pointing at the camera, suggests consoling cups of coffee, which everyone drinks in the style of a Japanese tea ceremony. More camera-pointing ensues. Dignity: 4/10

Madonna
Product: Takara sake (Japan)
Plot: The singer, in samurai garb, takes a half-hearted swipe at a gold CGI dragon, which disgorges a gold orb containing a glass of sake. She drinks it while her voice on the soundtrack wonders: "How can I be pure?" Dignity: 5/10

James Brown
Product: Nissin Cup Noodles (Japan)
Plot: Standing in someone's kitchen, the King of Soul sings an adapted version of *Sex Machine* while preparing a plastic pot of steaming "New!" miso soup. At least he looks like he's having fun. Dignity: 4/10

Arnold Schwarzenegger
Product: Alinamin V vitamin drinks (Japan)
Plot: Almost too bizarre to summarize. A goofy bespectacled Arnold gets into various awkward social situations, takes a furtive swig of Alinamin V and transforms into "Devil King V". Dignity: 3/10

SPICE MIXES OF THE WORLD

Over the past decade, thanks to authors like Yotam Ottolenghi and Sabrina Ghayour, Western home cooks have been exposed to a greater variety of cuisines from around the world – and with them a range of unfamiliar spice mixes, which have us reaching for Google to suss out their ingredients. Mixes vary, but here are some key ingredients in five of them.

MIXES

Berbere (Ethiopia and Eritrea) Add to slow-cooked wats (Ethiopian stews). Or sprinkle on grilled fish, burgers, meatballs, roast chicken.

Ras el hanout (North Africa) Use in stews and tagines, as a marinade or rub, or as a condiment.

Shichimi togarashi (Japan) An increasingly popular condiment in Japan. Sprinkle over rice, noodles, yakitori, tempura, steamed vegetables.

Jerk spice (Jamaica) Use as a rub or wet marinade for chicken, pork, goat or fish. Also goes with beef, lamb and vegetables.

Panch phoron (Bangladesh, Eastern India) Use in stews, as a rub for meats, sprinkled on vegetables or naan bread.

INGREDIENTS

Chilli powder, black pepper, fenugreek, cardamom, coriander seeds, ginger, cloves, cinnamon

Cardamom, turmeric, cloves, cinnamon, coriander seeds, nutmeg, mace, black pepper, cumin, paprika, allspice, ginger

Ground red chilli pepper, ground sansho, dried orange peel, sesame seeds, hemp seeds, ginger, garlic, shiso, nori

Allspice, Scotch bonnet pepper, cloves, cinnamon, nutmeg, thyme, garlic, ginger, sugar, salt

Cumin, fenugreek, fennel, black mustard, nigella in equal quantities

THE UBIQUITY OF CHICKEN

A subspecies of the Asian red jungle fowl (*Gallus gallus*), the domestic chicken is the most common type of poultry in the world. Though archaeologists believe it was first domesticated for cockfighting, it has become universally popular as food thanks to its mild taste and uniform texture, which translate to almost any cuisine.

Coq au vin (France)
Chicken + red wine + onions + lardons + mushrooms
A clever way of using a leathery rooster.

Chicken pastilla (Morocco)
Chicken + onions + almonds + spices + ouarka pastry + sugar
A dish for weddings and other special occasions.

Butter chicken (India)
Chicken + butter + spices + onions + cream + coriander
Said to have originated at Moti Mahal restaurant in Delhi.

Chicken mole (Mexico)
Chicken + chillies + chocolate + onions + coriander
Mole refers to a variety of sauces containing chilli.

Pollo alla cacciatora (Italy)
Chicken + onions + garlic + tomatoes + rosemary + wine
Aka hunter's chicken, a classic Italian dish.

Jerk chicken (Jamaica)
Chicken + allspice berries + Scotch bonnet chilli + lime
Traditionally served with rice and peas.

Chicken teriyaki (Japan)
Chicken + sake + mirin + soy sauce + sugar
Teriyaki sauce dates back to the 17th century.

Beggar's chicken (China)
Chicken + mushrooms + spices + chilli + rice wine
The chicken is stuffed, wrapped in mud and roasted.

Chicken tabaka (Georgia)
Chicken + garlic + cayenne pepper + herbs + salt
The chicken is flattened with a heavy weight, then pan-fried.

Cock-a-leekie (Scotland)
Chicken stock + leeks + prunes
Scotland's "national soup" may have originated in France.

ORIGINALITY IS OVERRATED

One thing most ambitious chefs strive for is to create something truly original. That's easier said than done, as even the most futuristic-seeming food relies on flavour combinations and techniques developed over millennia. **Corey Lee**'s response to this challenge, at his restaurant In Situ which opened at the San Francisco Museum of Modern Art in 2016, was to create a menu composed entirely of other chefs' dishes. By eschewing the pursuit of originality in his cooking, he created what the *New York Times* called "America's most original new restaurant". Here is the In Situ menu from March 2017, with credits to the original creators.

Sea Urchin in a Lobster Jell-o
Richard Ekkebus, Amber, Hong Kong, China, 2006

Crispy Crepe
Blaine Wetzel, The Willows Inn, Lummi Island, Washington, 2013

Lettuce Sandwich
Christian Puglisi, Relæ, Copenhagen, Denmark, 2015

Apocalypse Burger
Anthony Myint, Mission Street Food, San Francisco, 2016

Anis Marinated Salmon
Harald Wohlfahrt, Restaurant Schwarzwaldstube, Baiersbronn, Germany, 2010

Wasabi Lobster
Tim Raue, Restaurant Tim Raue, Berlin, Germany, 2013

Glazed Chicken Thigh
Hiroshi Sasaki, Gion Sasaki, Kyoto, Japan, 2015

The Voyage from the Indies
Olivier Roellinger, Les Maisons de Bricourt, Saint-Méloir-des-Ondes, France, 1982

Lamb Shank Mantı
Mehmet Gürs, Mikla, Istanbul, Turkey, 2012

The Forest
Mauro Colagreco, Mirazur, Menton, France, 2011

Kalbi Jjim
Roy Choi, L.A. Son, Los Angeles, California, 2013

Desserts

Oops! I Dropped the Lemon Tart
Massimo Bottura, Osteria Francescana, Modena, Italy, 2012

Jasper Hill Farm Cheesecake
Albert Adrià, Tickets, Barcelona, Spain, 2015

THE ULTIMATE TOMATO SAUCE?

For such a beautifully simple dish, there are endless disagreements over how to cook the perfect *sugo di pomodoro* to accompany pasta. Here are a few notable methods:

» **Marcella Hazan** simmers a tin of whole tomatoes in lots of butter with a peeled and halved onion, discarding the onion before tossing the sauce with the pasta.

» **Angela Hartnett** softens a finely chopped onion in olive oil, then adds garlic, tinned tomatoes, tomato purée, sugar and rosemary, and simmers until the sauce is thick.

» **Fergus Henderson** uses red onions, which he adds after the oil, garlic and tomatoes. (His tomatoes are chopped but not diced.)

» **Anna Del Conte** uses fresh tomatoes, brought to the boil with chopped celery, onion, carrot, garlic, parsley, sage, thyme and salt (adding tomato purée and sugar if the tomatoes are insipid).

For what it's worth, I keep it as simple as possible: oil and garlic, tinned whole tomatoes, simmered until completely reduced.

THE WORST KITCHEN
JOBS IN HISTORY

Before food processors and fan ovens, electricity and running water, many of the kitchen jobs that we consider quick and easy today – like roasting a chicken or beating egg whites – were time-consuming and back-breaking. If you were rich, you could avoid the worst of it by getting someone else, or a team of people, to do it all for you. But for the poor in centuries past, getting food on the table was a herculean task.

Turnspit – To cook meat in medieval England meant to roast it on a spit before an open fire. To ensure the meat cooked evenly, the heavy spits needed to turn, and for centuries the job of turning them was performed by young unfortunates (usually boys) crammed into small, hellishly hot spaces. Dogs and geese were also trained for the task before mechanization (mercifully) rendered them all obsolete.

Egg beater – During the Renaissance, it was discovered that eggs could be used as a raising agent in baking. The upside: cakes! The downside: hours spent beating egg whites into shape with highly inefficient pre-whisk technology – including spoons, knives, forks and bunches of birch twigs. Aching arms all round.

Grain crusher – Many staple foods such as wheat only become digestible when they are ground to a powder. Before the advent of mills, this arduous task had to be performed by humans, usually the women in a household, who would sit before a pestle and mortar, or a rudimentary stone quern, grinding away for hours on end.

Scullion – The lowliest servant in a household, the scullion (a male role later supplanted by the scullery maid) did all the jobs in the kitchen no one else wanted – cleaning pots and pans, peeling potatoes, plucking pheasants, scouring the floors – usually in the confines of a dank basement room.

Food taster – Not a kitchen position per se, but few would envy those who test a VIP's food to make sure it hasn't been spiked with poison. This archaic-seeming practice continues to this day (*see* page 56).

EATING FOR SPORT

It's pretty simple: you're given a pile of food and a time limit. You eat as much of the food as you can before the limit expires. Whoever consumes most wins. Competitive eating, an increasingly popular sport in North America and Japan, is often criticized for celebrating gluttony, but its stars treat it with utmost seriousness and wear their records – some listed below – with pride.

Joey "Jaws" Chestnut (USA)
Californian ranked #1 in the world by Major League Eating (MLE). Holds the Nathan's Hot Dog Eating Contest record (*see* page 116).
» Chicken wings (241 in 30 minutes)
» Matzoh balls (78 in 8 minutes)
» Hard-boiled eggs (141 in 8 minutes – beat that, Cool Hand Luke)

Sonya "The Black Widow" Thomas (USA)
Diminutive South Korean-born competitive eater, aka "The Leader of the Four Horsemen of the Esophagus", renowned for defeating men several times her size.
» Oysters (564 in 8 minutes)
» La Costena Jalapeno Peppers (250.5 in 9 minutes)
» Ears of sweetcorn (32.5 in 12 minutes)

Takeru "The Tsunami" Kobayashi (Japan)
Crashed onto the scene in 2001 when he doubled the Nathan's Hot Dog Eating Contest record. Has a trademark body wiggle called the "Kobayashi Shake", which helps get food down faster.
» Cow brains (57, or 8kg/7½lb, in 15 minutes)
» Pizza (62 slices in 12 minutes)
» Tacos (130 in 10 minutes)

Patrick "Deep Dish" Bertoletti (USA)
Mohawk-sporting Chicagoan, currently ranked #2 in the world. Drank 120 raw eggs on *America's Got Talent*.
» Cream-filled doughnuts (47 in 5 minutes)
» Pickled jalapenos (275 in 8 minutes)
» Gyoza (264 in 10 minutes)

WEIRD FEATS OF CONSUMPTION

What drives ordinary people to consume extraordinary things like billiard balls, bicycles and planes? In some cases it's straight-up showmanship: the compulsion to amaze for money and fame. In others, it's a genuine eating disorder called pica, which induces a craving for non-nutritive substances. And occasionally, it's a mix of the two, as pica sufferers turn this very strange ailment to their advantage.

An aeroplane – It took **Michel Lotito**, aka Monsieur Mangetout (1950–2007), just two years to eat an entire Cessna 150 plane, which would have weighed around 750kg (1,653lb). The French entertainer, who was diagnosed with pica, also ate shopping carts, televisions and 18 bicycles – nearly nine tons of metal over 40 years, by his own estimation – though curiously enough he was averse to bananas and hard-boiled eggs.

A shoe – In the 1970s, **Werner Herzog** told fellow filmmaker Errol Morris he'd eat his shoe if Morris ever completed and screened *Gates of Heaven*. The documentary came out to great acclaim in 1980, forcing Herzog to follow through on his promise. True to form, he made a short film about it, though he didn't eat the sole of the shoe, reasoning that one does not eat the bones of a chicken.

Light bulbs – As well as swallowing swords and knocking nails into his head, American showman **Todd Robbins** (b. 1958) is renowned for eating glass. After illuminating a light bulb to show that it works, he unscrews it, bites off the base and chews the glass like he's eating a piece of fruit. He estimates he's eaten up to 5,000 light bulbs over his career.

Poison – A great opponent of the Roman Republic, **Mithridates VI**, King of Pontus (135–63BC), was so paranoid about being assassinated that he consumed non-lethal doses of various poisons on a regular basis to immunize himself. So successful was his strategy that when he attempted suicide by poison after being defeated by Pompey, the attempt failed and he ordered his bodyguard to kill him with a sword instead.

YOUR COFFEE IS A CANVAS

There's more to latte art than the wonky heart traced with steamed milk on your espresso at the local coffee shop. Active since at least the 1980s (Jack Kelly at Seattle's Uptown Espresso was a pioneer), the practice is now the focus of competitions, notably the World Latte Art Championship, and countless Instagram feeds.

Pirate of the Caribbean 2
Elvis Matiejunas

Peacock
Caleb Cha

Einstein
Kohei Matsuno

Triceratops
Paulo Asi

Swan
Caleb Cha

Kangaroo
Elvis Matiejunas

Rose
Matthew Lakajev

Wolf
Ian Huang

Rosetta
James Hansen

SOME LIKE IT HOT

For many years, the king of the "superhots" – chillies that score above 500,000 on the Scoville scale – was the fiery red Savina pepper. The temperature increased in 2007 when the *Guinness Book of Records* acknowledged the Bhut Jolokia as the world's hottest (see my report below). Since then, the record has been smashed four times as superhots reach ever-more-fiendish levels of heat.

1994 – Red Savina
<u>Origin:</u> California, USA
<u>Scoville rating:</u> 570,000
<u>Effect:</u> "My first reaction was the same as anybody else's: I cried. It's a very intense heat, it starts very far back in the throat and when that heat hits you, it'll bring you to your knees." — Jim Campbell, *Food Network*, 2002

2007 – Bhut Jolokia (aka the Ghost Chilli)
<u>Origin:</u> Northeast India
<u>Scoville rating:</u> 1,001,300
<u>Effect:</u> "For some reason, I thought I would be able to weather the chilli with dignity. Instead, I became a storm of flailing limbs and strangled protests… The worst of it was over within 20 minutes, but for the next few hours I wondered if my mouth and head would ever feel the same." – Killian Fox, *The Times*, 2008

2011 – Infinity Chilli
<u>Origin:</u> Lincolnshire, England
<u>Scoville rating:</u> 1,176,182
<u>Effect:</u> "The fire hits my throat and then goes into my ears. My ears! They ache terribly. My legs wobble. Punch-drunk, I slump on a garden chair… After five minutes, the heat subsides to phall level. The ear agony goes. Now I can feel it in my stomach. But I feel good too: high and dizzy, as if drunk." – Patrick Barkham, *The Guardian*, 2010

2011 – Naga Viper
<u>Origin:</u> Cumbria, England
<u>Scoville rating:</u> 1,382,118

Effect: Simon Jack: "Okay, here goes my first bite [coughs]. At the moment, it's tolerable. But there's an angriness to it... which is getting more intense [pauses and clears breath a few times] making it quite hard to breathe. I might pass it over to Tom here for a moment. I just... for the challenge I'm not allowed any water, right?" Tom: "No. Okay, so what's happening here is increased heart rate, change of skin complexion; so he's now gone bright red, finding it hard to breathe. I think the hiccups are probably coming on. Eyes starting to stream, and it looks really quite excruciating to be honest with you." Simon: "Um..." Tom: "And a fair amount of distress." Simon: "Um..." Tom: "Where is it burning?" Simon: "The back of my tongue. I've got a headache now. I've had enough now." – Simon Jack, *BBC World Service*, 2015

2011 ~ Trinidad Scorpion Butch T

Origin: Morisset, Australia

Scoville rating: 1,463,700

Effect: "At first, the pepper didn't taste hot at all – in fact, it had a gentle floral flavour. After a few seconds, the heat began to hit. This pepper was hotter than anything I had ever tasted. But it was about to get worse. The scorpion pepper creeps up on you, getting incrementally fiercer... until your whole face feels like it has turned into lava."
– Jackson Landers, *Slate*, 2013

2013 ~ Smokin' Ed's Carolina Reaper

Origin: South Carolina, USA

Scoville rating: 1,569,300

Effect: "All hell just broke loose inside my mouth. My tongue is burning. My upper lip is stinging. My eyes are bloodshot. It's like being face-fucked by Satan himself... Without warning a numbness shoots through my right pinkie, then up into my biceps. Strangely, a mellow head rush sets in. My pupils dilate as a tear trickles down my cheek."
– Steven Leckart, *Maxim*, 2013

THE RULES OF ATTRACTION

If you're the kind of cook who finds it difficult to make anything vaguely elaborate without referring to a recipe book, track down **The Flavour Thesaurus** by Niki Segnit. This ingenious tome lists 99 ingredients, divided into 16 categories such as marine, citrussy and earthy, and suggests which ingredients work well together. There are recipes thrown in, but in general the book encourages freestyling and gives you a framework to do so with confidence. Some of the 980 recommended pairings, like beef & horseradish, seem obvious; others (at least to my northern European palate) are less so. Have you ever tried combining any of these?

Pineapple & anchovy	Banana & caviar	Goats' cheese & blackberry	Mushroom & oyster
Bacon & anise	Grape & white fish	Mango & shellfish	Orange & beef
Avocado & coffee	Caviar & white chocolate	Coffee & coriander seed	Watermelon & pork
Cumin & mango	Cherry & smoked fish	Bacon & chocolate	Cauliflower & chocolate

Of the 99 flavours listed, the most sociable are **pork** (48 pairings), **potato** (44), **chicken** and **almond** (40 each); while the least outgoing are **swede** (7 pairings), **washed-rind cheese, blackberry** and **blackcurrant** (8 each).

ISINGLASS

In 2015, Guinness announced that they were removing a fish product from their production process. Now the company's stout on draught (though not yet in bottles or cans) is free of isinglass, a form of collagen that comes from the dried swim bladders of fish such as hake, cod and catfish. Other brands, including Foster's and Carling in the UK, still use isinglass to clarify their beers – and it also features in wine-making and confectionery.

THE WACKIEST FOOD INVENTIONS

In today's lazy lazy world, there's no inconvenience so minor that some enterprising soul hasn't sought to alleviate it with an ingenious and wholly unnecessary fix. Tired of stirring your porridge in the morning? The Stirio Automatic Rechargeable Pot Stirrer will do it for you. Infuriated with butter that refuses to spread? Buy a self-heating butter knife. Here are some other food inventions you may want but almost certainly don't need.

What is it?	What does it do?
Twirling Spaghetti Fork[1]	Saves you from the hassle of wrapping spaghetti around your fork
Portable Pizza Pouch	Stores a single slice of pizza in a plastic holder worn around the neck
Keyboard Waffle Iron	Makes a waffle in the shape of a computer keyboard
Selfie Spoon	Allows you to take hands-free photographs of yourself while eating
Flask Tie[2]	Lets you drink at work from a pouch which fits neatly inside your neckwear
Anti-Theft Lunch Bags	Deters sandwich theft by means of fake mould spots on see-through plastic
Forkchops	Combines a fork and chopsticks in one handy utensil
Giant Wine Glass	Holds an entire bottle of wine in a single glass
Fondoodler	Lets you scribble on surfaces with molten cheese
Selfie Toaster	Prints images of your face onto hot bread
Banana Surprise Yumstation	Allows you to burrow a hole in a banana and pump it with fillings of your choice

[1] See also: Motorized Ice Cream Cone, which saves you from the inconvenience of rotating the cone by hand. [2] See also: Wine Rack, a bra with pouches and a drinking tube.

THE FIVE BASIC TYPES
OF ASTRONAUT FOOD

Rehydratable – Foods that have had water removed by sublimation or freeze-drying to reduce weight, or dried foods such as cereals.

Thermostabilized – Canned foods, including standard goods such as peas, tuna and chocolate pudding.

Intermediate moisture – Low-moisture foods such as dried peaches, apricots or beef, packed in plastic.

Natural form – Fresh foods, nuts, granola bars, bread.

Beverages – Powdered drinks work best in space. The only natural juice currently used is freeze-dried orange and grapefruit juice.

A BRIEF HISTORY OF FOOD IN SPACE

Over the millennia, humans have improved at packing non-perishables into cramped spaces for long journeys across inhospitable regions of the earth, such as oceans and deserts. In the last 70 years, however, with the advent of space travel, we've had to step up efforts to sustain travellers far from home – with food that's extremely lightweight and compact as well as nutritious and at least tolerably tasty.

1961 – The first man in space, Soviet cosmonaut **Yuri Gagarin**, eats the first meal in space on 12 April. On the menu: tubes of beef and liver paste, with chocolate sauce for dessert. His colleague Gherman Titov, on the second-ever manned space mission four months later, becomes the first person to vomit in space.

ГОВЯДИНА   BEEF

ВОДКА   VODKA

1962 – The first American in space **John Glenn** discovers (presumably to his relief) that microgravity does not affect the swallowing process. He consumes a tube of apple sauce.

1965 – Gemini 7 astronauts **Frank Borman** and **Jim Lovell** are allowed just 1.7lb (771g) of food per day for their 14-day mission into space. Today's shuttle crews are allowed 3.8lb (1.7kg) of food per day.

1969 – The first meal on the moon is eaten by **Neil Armstrong** and **Buzz Aldrin** on 20 June. They chow down on bacon squares, peaches, sugar cookie cubes, pineapple grapefruit drink and coffee.

1975 – Astronauts and cosmonauts dine together during the Apollo-Soyuz Test Project (1975). The Russians provide canned beef tongue, packaged Riga bread and tubes of borscht labelled "vodka".

1985 – Astronaut **Mary Cleave** and payload specialist **Rodolfo Vela**, the first Mexican astronaut, introduce tortillas to NASA's shuttle menu. Well-suited for space cuisine, tortillas produce few crumbs and, with reduced water and oxygen contact, can last for several years. They also make a very serviceable Frisbee.

1995 – US astronaut **Bill Gregory** eats shrimp cocktail over 48 consecutive space meals. This 1950s mainstay is regularly cited by US astronauts as their favourite extra-terrestrial treat.

2003 – For China's first manned space flight in October, astronaut **Yang Liwei** dines on specially processed *yuxiang* pork, *kung pao* chicken and eight treasures rice, along with Chinese herbal tea.

2008 – South Korea's first astronaut, **Yi So-yeon**, a crew member on the International Space Station, brings modified kimchi into space, after three research institutes spend several years and over $1m creating a version of the fermented cabbage dish suitable for space travel.

2015 – On 3 May, Italian astronaut **Samantha Cristoforetti** becomes the first person to drink freshly brewed coffee in space with the help of the ISSpresso machine developed by Lavazza and Argotec.

UNUSUAL COFFEE CONCOCTIONS

Milk and sugar is one thing; it's quite another to put eggs, cheese or nitrogen gas in your daily cup of coffee. Here are some of the more unusual combinations you might encounter in coffee houses and other sites of caffeine consumption around the world.

Kaffeost (Sweden/Finland)
Leipäjuusto cheese + black coffee
The cheese can be dipped or immersed in the coffee.

Coffee & tonic (Sweden)
Ice + tonic water + espresso
Invigorating drink from the Koppi coffee shop in Helsingborg.

Chicory coffee (USA)
Coffee + ground root chicory
Chicory has been used as a additive or substitute during coffee shortages.

Nitro coffee (USA)
Cold-brewed coffee + nitrogen gas
Developed by Stumptown Coffee Roasters in Portland, Oregon, later adopted by Starbucks. The nitrogen provides a rich mouthfeel and sweet flavour.

Cà Phê Trung (Vietnam)
Egg yolk + condensed milk + black coffee
A speciality of Hanoi. Try it at Café Giang.

COFFEE FIENDS

The links between coffee and creativity are not mysterious. For centuries, artistic types have relied on caffeine to fuel their work, some treating its stimulating powers with great respect, others guzzling insane quantities or even literally eating the stuff to keep themselves going for an extra few hours. How different our creative landscape today would be without the perky coffee bean.

» "Coffee is a great power in my life," wrote **Honoré de Balzac** in 1830; "I have observed its effects on an epic scale." Indeed he did. When in the grip of one of his "orgies of work", the French novelist and playwright would get up at 1am and write till 4pm with a 90-minute nap in the middle. To fuel himself, he imbibed as many as 50 cups of coffee a day. He also dabbled with "a horrible, rather brutal method" which involved eating pure coffee grounds on an empty stomach. When he did this, he wrote, "Ideas quick-march into motion like battalions of a grand army to its legendary fighting ground, and the battle rages." For Balzac, the battle raged until his death at 51: he wrote 91 long and short works of fiction in the space of just 16 years.

» In her 1988 novel *Cat's Eye*, **Margaret Atwood** writes: "I don't even glance at the herbal teas, I go straight for the real, vile coffee. Jitter in a cup. It cheers me up to know I'll soon be so tense." The Canadian author is a committed coffee drinker in real life – she created her own blend in 2012 with Balzac's Coffee Roasters in Toronto. Quite unlike the vile, jittery substance from *Cat's Eye*, the Atwood Blend – still available today – is "a mild and gracious cup with a supple, caramel finish".

» According to **Ludwig van Beethoven**'s secretary Anton Schindler, the great composer would assign no more or no fewer than 60 beans to his morning cup, often counting them out by hand to ensure he was getting the exact dose.

» **David Lynch** claims his relationship with coffee began at the age of three. At one stage the filmmaker was drinking 20 cups a day; nowadays he averages 10, though the cup size has increased. A good coffee, he says, "should have no bitterness, and it should be smooth and rich in flavour. I like to drink espresso with milk, like a latte or a cappuccino, but the espresso should have a golden foam. It can be so beautiful."

» The Danish philosopher **Søren Kierkegaard** took coffee with his sugar rather than the other way round, according to his biographer Joakim Garff. "Delightedly he seized hold of the bag containing the sugar and poured sugar into the coffee cup until it was piled up above the rim. Next came the incredibly strong, black coffee, which slowly dissolved the white pyramid."

TEN COOKING TIPS
FROM THE GANNET

The most common piece of kitchen wisdom we've been given, over the course of our interviews for *The Gannet*, is to clean up as you go along. ("A tidy kitchen is a good thing," the beloved London chef Jeremy Lee told us, recommending "A little bit of order – you don't have to be martinet about it.") Here are some other sage bits of advice we've gathered along the way.

1 "In case you end up adding too much salt to *any* dish, add a piece of raw peeled potato to the dish and let it cook for a bit. The potato will absorb the excess salt." – **Asma Khan, chef, London, UK**

2 "Always have a white vinegar in your kitchen – you can clean everything with it and it doesn't contain any chemicals – I use it instead of bleach. And use bicarbonate of soda to clean your pots and pans if they're all burnt. Even better, mix the two together." – **Alix Lacloche, food writer, Paris, France**

3 "I love to slow-cook, which means that you put something in the oven and you wait for two, four, six, eight hours, and during this time you can read a book or go for a swim. This is something I really like – slowing down." – **Hervé Tullet, children's author, Paris, France**

4 "Read the recipe. A lot of people don't, and then they get home and they're like, 'I need eggs for this?', and they're mad and have a bad time and end up ordering takeout. But also, feel free to ignore the recipe! If you hate coriander [or cilantro], replace it with something else. If you love cumin, add cumin. It's not always about making a perfect plate of food; it's about empowering yourself to do your own thing." – **Alison Roman, food writer, New York, USA**

5 "If you're feeling uninspired, pull out an onion, get a knife and chop it up – that's the best way to get over the hump, that feeling that you don't know what to cook. Once you start chopping, the rest will come to you. I don't know if it's the sensory thing from the smell, but that's really the basis of so many things." – **Addie Broyles, food journalist, Austin, TX, USA**

6 "Never wash fish. Some recipes tell you to rinse fish, but the French would put you in an asylum. They say that all the micro-organisms on the skin of the fish have so much flavour, so if you wash them, you're washing away a whole bunch of the taste. On a more philosophical note, there was a wonderful chef in Brittany who once said to me, 'Alec, if you calm your mind down, the product always tells you how to cook it', and I think that's really true." – **Alec Lobrano, restaurant critic, Paris, France**

7 "If you leave onions in a pot for five minutes, you'll have an okay result. But if you leave them to cook really slowly in butter and bay leaves for 20 minutes, you'll have a dream. It's all about patience. My whole approach to cooking is about doing things slowly." – **Paul Flynn, chef, Waterford, Ireland**

8 "I think water is the forgotten ingredient in the kitchen. Often students [at Ballymaloe Cookery School] find they've made something too intense and they don't know what to do. Obviously I tell them taste, taste, taste – always – but water is the thing that will reduce the concentration, and of course it has to be good quality. A pasta sauce that's too concentrated might taste delicious in one spoonful, but you won't enjoy a full plate of it, and water can fix that." – **Darina Allen, food writer and educator, Cork, Ireland**

9 "Cooking is mostly chopping stuff up and making it hot. The heat is important. A lot of people roast vegetables at 350°F (180°C/gas mark 4) when they should really put it at 450°F (230°C/gas mark 8) and give some character to their food. When you're sautéing something in a pan, use the fire – use it until it's too much and then turn it down. People are very timid when it comes to using heat and their cooking suffers – you need to make those chemical reactions happen and give the food some character."
– **David Ansel, soup entrepreneur, Austin, TX, USA**

10 "Don't panic. What could possibly go really, really wrong? It's not life-threatening." – **Anna Koska, food illustrator, East Sussex, UK**

FOODS TO COOL THE HEAT OF LUST

Most of us are familiar with the concept of aphrodisiacs, if not their actual effect. On cuneiform tablets from 800BC Babylon, it was written that by cutting the head off a partridge, eating its heart and draining its blood into a cup of water, leaving the mixture overnight and drinking it the next morning, you would give your libido a much-needed boost. More recently, and less gruesomely, we attribute aphrodisiac qualities to oysters, chocolate and Champagne. But what about foods that have the opposite effect? It turns out that anaphrodisiacs, or libido-dampeners, have a long and colourful (if scientifically dubious) history too…

Chasteberry – The Roman naturalist Pliny the Elder believed the stems and leaves of this Mediterranean shrub were used to "cool the heat of lust" in Athens during the Thesmophoria, a festival for which women left their husbands' beds and swore off sex.

Alcohol – Though it can increase sexual desire in the short term by reducing inhibitions, studies have shown that alcohol decreases arousal over time and may inhibit testosterone production in men.

Liquorice – Its root has been shown to lower testosterone in men and decrease libido. Of all the purported anaphrodisiacs, this is perhaps the most likely to have the desired effect.

Marjoram – This and various other herbs, including coriander, have long been promoted as anaphrodisiacs, though there is little evidence that they have any effect.

JOOK-SING

These rare Cantonese noodles are traditionally made with duck eggs and pressed in an extraordinary way: a bamboo log is laid across sheets of dough on a counter and wedged tightly at one end; the noodle-maker sits on the other end of the log and see-saws up and down on the sheets – to give the final product extra spring. They are still made at Kwan Kee restaurant in Hong Kong.

THE AMUSEMENT FACTOR

Sometimes it's too much bother trekking around in search of interesting food: you want it all in the one place with added rollercoasters. The food theme park, complete with rides and interactive "experiences", is a relatively recent phenomenon – the first ones emerged in Japan in the 1990s. Here are four interesting examples.

Shin-Yokohama Raumen Museum (Yokohama, Japan)
Established: 1994 Theme: Ramen
Admission (adult): 310 yen (£2.10/$2.80)
Attractions: A museum exploring the history of Japan's favourite noodle soup (which originally came from China). Two floors containing a replica of Tokyo streets in the 1950s when ramen was taking off, including nine restaurants serving regional variations.

FICO Eataly World (Bologna, Italy)
Established: 2017 Theme: Italian food (Eataly is a chain of Italian marketplaces) Admission: Free
Attractions: Six rides, 40 workshops (for making cheese, pasta and so on), 25 restaurants and stalls, and 500 adult-sized Bianchi tricycles with shopping baskets for scooting around the 20-acre, $106m park.

Cité du Vin (Bordeaux, France)
Established: 2016 Theme: Wine
Admission (adult): €20 (£16.80/$23.60)
Attractions: Twenty themed areas and exhibits, including an overview of world wines and a boat ride simulator. Three hi-tech tasting rooms, a shop with 800 wines (just 200 of which are from France). A rooftop restaurant and bar.

Tayto Park (Meath, Ireland)
Established: 2010 Theme: Potato crisps (Tayto is Ireland's largest manufacturer) Admission (adult): €30 (£25.20/$35.40)
Attractions: Ireland's only wooden rollercoaster. A zoo. A zip line. A bewildering amount of Tayto merchandise. All visitors to the 55-acre park receive a complimentary bag of crisps when they leave.

THE ULTIMATE BOILED EGG?

Few culinary tasks are simpler than boiling an egg. And yet, despite the short list of components – pan, water, egg, heat – there are various methods, each of which can be tweaked in many tiny ways (and some of which do away entirely with pan and water). Here are four prominent approaches to the humble boiled egg. (All timings are for soft-boiled, unless otherwise noted.)

» **Gordon Ramsay** favours the hot start: he brings a pan of water to the boil, lowers the eggs in gently on a spoon, returns the water to the boil and counts to five, then turns the heat down and cooks the eggs for 4½ minutes.

» **Martha Stewart** favours the cold start: she places the eggs in a saucepan large enough to accommodate them in a single layer, fills the pan with cold water, covering the eggs by an inch, and brings the water to a boil over a medium–high heat. She then turns off the heat, covers the pan and lets it stand for 1½–2 minutes before removing the eggs from the water and serving immediately.

» **J Kenji López-Alt**, author of *The Food Lab*, recommends steaming instead of boiling: the eggs cook more gently; you don't have to worry about lowering the temperature; and you only have to boil an inch of water, speeding up the process. Once the water is boiling, he adds the eggs to the steamer, covers it and cooks for 6 minutes (or 12 for hard-boiled).

» French cookery scientist **Hervé This** says cooking eggs is a question of temperature, not time. For a soft yolk, he heats his oven to exactly 65°C[1] and simply places the eggs inside for about an hour (they can stay in longer). For a firmer egg, he will raise the temperature as high as 80°C and cook it for the same amount of time.

[1] For this to work, you need an accurate oven (most domestic ovens deviate by as much as 10°C) and it probably doesn't make sense to waste all that energy for just one or two eggs. The resulting eggs, however, are pretty special.

THE GOURMAND'S GALLERY

The history of art is a cornucopia of culinary delights. It abounds with tempting fruits and fleshy meats, scarlet lobsters and ruby-coloured wine, much of it freighted with symbolic intent. Here are eight paintings and one piece of performance art in which food is front and centre. Can you make them out?

1) Thirteen men cluster around a long table, six to each side of a central figure who stares down at the plates and bread rolls (1498)

2) A man with a wicker hat and a face made of fruit stares from the frame; invert the painting and he becomes a fruit basket (around 1590)

3) Two men carry a tray laden with soup and porridge through a countryside banquet while another man fills a jug with wine (1556)

4) A businessman in suit and bowler hat faces the viewer, his features obscured by a leafy green apple (1964)

5) Two men and a woman sit at the counter of a late-night New York diner, attended to by a man in white uniform (1942)

6) A woman in a white headdress pours milk from a jug into a bowl (around 1660)

7) Two clothed men and a naked woman recline in a wood while their picnic lies in disarray beside them (1863)

8) A husband and wife serve a huge roast turkey to a table of grinning guests (1943)

9) A woman scrubs thousands of bloody beef bones (1997)

1) *The Last Supper*, Leonardo da Vinci 2) *Reversible Head with Basket of Fruit*, Giuseppe Arcimboldo 3) *Peasant Wedding Feast*, Pieter Bruegel 4) *The Son of Man*, René Magritte 5) *Nighthawks*, Edward Hopper 6) *The Milkmaid*, Johannes Vermeer 7) *Le Déjeuner sur l'Herbe*, Édouard Manet 8) *Freedom From Want*, Norman Rockwell 9) *Balkan Baroque*, Marina Abramović

MONKS MAKING DRINKS

When Christ turned water into wine at the wedding at Cana, he was setting an example that his followers would pursue with gusto in centuries to come. After the Rule of St Benedict (480–547), which stated that monks should work to support themselves, monasteries in Europe developed a mastery of brewing and wine-making. Sale of alcoholic beverages remains an important source of income for some orders today.

Chartreuse – This vivid green liqueur flavoured with 130 herbs and other plants has been distilled by Carthusian monks in southeast France since 1737. The secret recipe is said to derive from an alchemical manuscript presented to the monks in 1605.

Champagne – The French Benedictine monk Dom Pérignon (1638–1715) didn't invent Champagne – he actively laboured to remove its bubbles, which were seen as a flaw at the time – but his pioneering techniques did influence its development.

Buckfast – Benedictine monks at Buckfast Abbey in Devon first made this caffeinated fortified wine in 1890. Originally sold as a medicine, it has more recently been linked with antisocial behaviour in Scotland.

Trappist beer – Six monasteries in Belgium, five around Europe and one in the US, are accredited to brew and sell this highly regarded (and often high-alcohol) beer. Westvleteren 12, brewed at the Trappist Abbey of Saint Sixtus in Vleteren, Belgium, often ranks as one of the world's best beers (*see* page 12).

In the Buddhist traditions of Asia, meanwhile, monks are often associated with the production of **tea**. Monasteries and temples in China, and later Japan and Korea, had a strong influence on how tea was grown, processed and served. With its calming, cleansing qualities, not to mention the meditative aspects of brewing, tea seems a natural fit with Buddhism (*see* page 46) – much more so than booze.

FERMENTATION IS EVERYWHERE

It's become a big food craze in recent years, but fermenting is absolutely not a niche activity. "Everybody is eating products of fermentation every day," says Sandor Katz, American author of *The Art of Fermentation*. Many staples of our daily diet are fermented; here's a small selection of them, alongside some more challenging examples.

BEGINNER

Chocolate – When cocoa pods are processed, they are first fermented for 5–8 days in a warm environment to break down the pulp around the beans and develop flavour.

Yogurt – Bacterial cultures are added to milk, metabolizing certain compounds to produce yogurt's texture and characteristic tartness.

Bread – Any bread that contains raising agents like yeast and is given time to rise undergoes fermentation.

Coffee – All coffee is fermented, but methods differ. Most commonly, coffee cherries are pulped, removing the flesh, then the beans are fermented in water tanks and dried.

Wine – Without fermentation, there would be no alcohol. Wine takes shape when yeast converts the sugars in grape juice into ethanol.

ADVANCED

Natto – Smelly, strong-flavoured, stringy and slimy, Japanese fermented soybeans are not to everyone's taste (though Katz, who has tried *natto*, says he "really loves" it).

Hákarl – A gutted and beheaded shark is buried for 6–12 weeks, then hung in the shade for another few months. The taste, say those brave enough to try this Icelandic specialty, is better than the smell.

Kombucha – Sugared tea is added to a slippery, translucent SCOBY (symbiotic colony of bacteria and yeast) and left to ferment until it becomes lightly effervescent.

Stinky tofu – Mostly eaten in outdoor settings, perhaps due to the fiercely strong aroma, this Chinese delicacy is usually served deep-fried with a spicy or sweet sauce.

CONDIMENTS OF THE AMERICAS

i. BBQ sauce – Commonly associated with the Deep South and big throughout North America. Maple syrup is sometimes added in Canada.

ii. Mayonnaise – The USA consumes $2bn-worth a year and the market for low-fat mayo doubled between 2005 and 2014.

iii. Ketchup – Heinz claimed 61.5 per cent of the market in 2015 in the country that gave the world burgers and hot dogs.

iv. Mustard – Another fast-food classic. To many Americans, a squeezy yellow bottle on a diner counter or hot-dog stand.

v. Pico de gallo – Aka *salsa fresca*, this Mexican classic is typically found in tacos, fajitas, salads and on egg dishes.

vi. Hot sauce – Marie Sharp's "Belizean Heat" bottles have been gracing tables throughout the country since 1980.

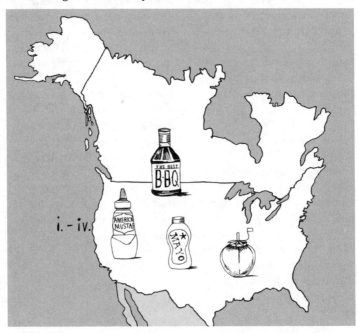

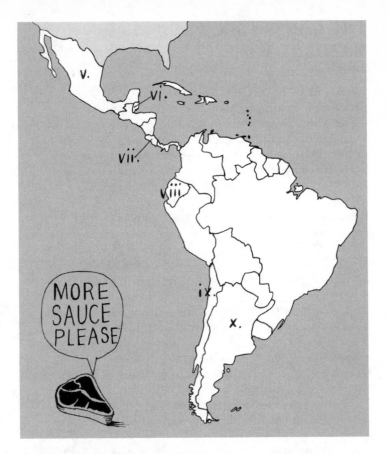

vii. Salsa Lizano – This Costa Rican light brown sauce was developed in 1920 and is used with tamales and *gallo pinto* (rice and black beans).

viii. Salsa di mani – In Ecuador, this peanut sauce is used as a topping for potato patties, vegetables and meats.

ix. Pebre – Ubiquitous in Chile, this sauce varies by place but usually contains coriander, garlic, onion, olive oil and ground aji chillies.

x. Chimichurri – A green or red sauce containing oil, vinegar, garlic and other herbs, spices and vegetables. Especially popular in Argentina, where it typically goes with grilled meats.

THE GANNET EXPLAINS...
NATURAL WINE

What is it?

Natural wine is a tricky term which irritates some as much as it excites others. It refers to wines that have been grown and processed with minimal intervention: no chemical additions in the vineyard or the cellar. These wines tend to be produced in small quantities by independent growers with the aim of expressing terroir – an unmediated sense of the place where the vines are rooted.

What chemical additions are we talking about?

The big one is SO2, which occurs naturally during fermentation, but which is added in large quantities by many conventional winemakers to stabilize their product. Most natural winemakers add little or no SO2 (aka sulphites), believing that it mutes the wine's nuances and may even be damaging to health. They also steer clear of dozens of other additives, from sugar and yeast to tartaric acid and isinglass (fish bladders, *see* page 74), permitted in wine production.

Sounds like a good thing. What's the problem?

One issue is with the name itself, which implies that all other forms of wine-making are unnatural. Another is with the lack of accreditation: any chancer can call their wines "natural" – and charge extra for it – without being held to account by a regulator.

Okay I can see why that's an issue. But what about the taste?

It's hard to generalize – natural wines vary no less than conventional wines, perhaps more so. But there are certain divisive characteristics: the cidery funk of some whites, the vinegary tang and farmyard aromatics of certain reds. Critics hold these to be faults, the results of undisciplined or inept wine-making. It's true that, by relying on fewer additives, natural wine-makers expose themselves to greater risks. Things can go badly wrong and there is less consistency. But this, a proponent would argue, is part of the excitement: it's worth it for the vibrancy and character of a good natural wine. One thing that's for certain is that it's not going away: *Decanter* magazine called the rise of natural wine "the major wine development of the century so far" – and that movement is only going to get bigger.

NORDIC FOOD LAB'S "BUG FOLIO"

Established in 2008 as a research offshoot of Copenhagen restaurant Noma, the Nordic Food Lab has expanded far beyond its original remit. One of its big preoccupations is to make insects more appealing to Western palates – and if you venture into the Bug Folio section of its website you'll find a trove of weird and wonderful ento-recipes, including these.

Anty gin & tonic	Cricket broth
Desert locust tabbouleh	Dung beetle larvae
Grasshopper garum	Roasted termite queen
Arthropod surf 'n' turf	Hornet highball (Japanese
Bee larvae mayonnaise	Witjuti grub and watermelon
(shrimp and crickets)	giant hornet liquor, whisky,
Moth mousse	soda, lemon)
Peas 'n' bees	

One recipe that jumped out at me from the Bug Folio was "Ants on a log". The recipe is based on a classic snack of North American children from the 1950s, involving raisins (or "ants") on a stick of celery coated in peanut butter (the "log"). "The metaphor was too tempting not to make literal", state the authors, and in their version, the log is made with celeriac, butter and birch bud salt, which is then rolled in a crumb mix of celeriac peel, ground coffee and dark rye bread. Onto this are placed at least 40 smelling carpenter ants – a big surprise for any nostalgic American baby-boomer expecting raisins.

season to taste

LUNCHBOXES OF THE WORLD

Bento box (Japan)

Bento ("convenient" in Japanese) is not a recent concept: these neatly sectioned lunchboxes date back at least to the late Kamakura period (1185–1333) when packing cooked, dried rice for lunch became popular. Traditional wooden lacquer boxes have now mostly given way to plastic containers available to buy in shops and train stations, though it's still common to prepare bento boxes at home and bring them to school or work.

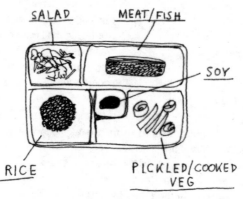

SALAD MEAT/FISH
SOY
RICE PICKLED/COOKED
VEG

Tiffin box (India)

An Indian invention, though popular as far away as Indonesia and Hungary, these ingenious stacked containers (also known as *dabbas*) are usually made of steel, which keeps food warm(ish) until lunchtime. Most have two or three layers, though fancy ones have four. In Mumbai, a super-efficient network of *dabbawalas*, or lunchbox carriers, transport boxes to city office workers from suburban homes.

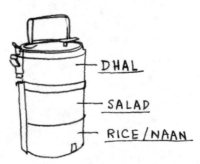

DHAL

SALAD

RICE/NAAN

Packed lunch (Ireland)

Relatively unsophisticated compared to its counterparts in Japan and India, the Irish lunchbox of my childhood almost always centred around a sandwich and a piece of fruit, with a carton of juice on the side. My mother took a firm stance on sugary food (or "rubbish", as she called it), so I could only envy the chocolate bars and crisp packets that cropped up in the lunchboxes of my peers.

SANDWICH   APPLE

THE WORLD'S BEST BAKERS

In February 2016, the hosts of the Coupe du Monde de la Boulangerie, France – aka the forefathers of baking – suffered a serious upset when Taiwan and South Korea edged ahead of them on the medals table. The South Korean winners impressed judges with challah bread, Swedish *kanelbulle* (cinnamon rolls), and an archer on horseback made entirely of bread, with savoury pastries and mini-burgers arrayed around his feet. The victory followed the 2014 opening of a huge South Korean bakery chain named Paris Baguette, in the heart of the French capital.

KANGATARIAN

A person (almost certainly living in Australia) who avoids all types of meat apart from that of the kangaroo – the ethical and environmental justifications being that kangaroos roam free, are usually killed humanely with a single shot to the head and emit much less methane than domestic livestock.

REVEALING THE LAYERS

Why go to the trouble of assembling a dish layer by painstaking layer, instead of mixing it all in together like a stew? In some cases, the integrity of the dish depends on it (try imagining a layerless sandwich). In others, it's for purely aesthetic reasons and the pleasure of cutting in to reveal the stratified interior, like a geological formation built up over millions of years rather than an hour or two in the kitchen on a wet Saturday afternoon. Can you guess what these layered dishes are?

Savoury

1

Basil
Mozzarella
Grated Parmesan
Pasta sheets
Ragu
Pasta sheets
White sauce
Pasta sheets
Ragu

2

Bun
Mayo
Lettuce
Tomato
Onions
Bacon
Ketchup
Cheese
Burger
Bun

3

Bread
Mayo
Lettuce
Tomato
Bacon
Bread
Tomato
Chicken/Turkey
Lettuce
Mayo
Bread

Sweet

4

Almonds
Whipped cream
Custard
Amaretti
Sherry
Fruit compote
Boudoir biscuits

5

Cocoa
Mascarpone mix
Dark rum
Savoiardi biscuits

6

Chocolate
Fondant
Puff pastry
Crème pâtissière
Puff pastry
Crème pâtissière
Puff pastry

1) Lasagne, 2) Cheeseburger, 3) Club sandwich, 4) Sherry trifle,
5) Tiramisu, 6) Millefeuille

THE GREATEST TOUR RIDER?

Before a musician of a certain stature plays a gig, their management sends the venue a "rider" – a list of backstage food and drink requirements that often veer into fabulous excess. There is an archive of these on *The Smoking Gun* website, but **Grace Jones**'s glorious Champagne-and-oyster-fest, cited in her 2015 book, *I'll Never Write My Memoirs*, stands out.

6 Bottles of Louis Roederer Cristal Champagne
3 Bottles of French Vintage red wine (e.g. St Emilion, Medoc, Bordeaux)
3 Bottles of French Vintage white wine (e.g. Sancerre, Pouilly Fuisse)
2 Dozen Findeclare [sic] or Colchester oysters on ice (unopened) –
(Grace does her own shucking.)
2 Sashimi and Sushi platters for 8 people
6 Fresh lemons
1 Bottle of Tabasco sauce
1 Fresh fruit platter for 8 people
6 Bottles of Coca Cola
12 Bottles of still and sparkling water
12 Bottles of fresh fruit juices
Wine glasses, Champagne flutes, tumblers (all glass, no plastic)
Cutlery and sharp knife (include 1 oyster knife)
1 Make-up mirror (no neon strip lighting, only opaque white bulbs)
3–4 Bunches of flowers – prefer lilys [sic] and orchids

OTHER EYE-CATCHING RIDER DEMANDS

Mötley Crüe's rider in the 1980s asked for local AA meeting schedules, a sub-machine gun, a 12-foot boa constrictor and a jar of Grey Poupon mustard.

On his *Saint Pablo* tour, **Kanye West** demanded a slushy machine with mixes of Coke and Hennessy, and Grey Goose and lemonade.

Axl Rose asked for "wines, beers, vodka, red and white roses and a square melon" – an "exotic" item "that absolutely can't be missed". He also requested a custom-made Italian leather couch.

JACK WHITE'S GUACAMOLE

In February 2015 the University of Oklahoma's student newspaper published a copy of the former White Stripes frontman's rider prior to his gig on campus. It warned against including any bananas with the words: "This is NO-BANANA TOUR. (Seriously.) We don't want to see bananas anywhere in the building." It also included a detailed recipe for guacamole devised by his tour manager, Lalo Medina:

8 large, ripe Hass avocados (cut in half the long way, remove the pit – SAVE THE PIT, THOUGH – and dice into large cubes with a butter knife. 3 or 4 slits down, 3 or 4 across. You'll scoop out the chunks with a spoon, careful to maintain the avocados in fairly large chunks.)

4 vine-ripened tomatoes (diced)

½ yellow onion (finely chopped)

1 full bunch of coriander (chopped)

4 Serrano peppers (de-veined and chopped)

1 lime

Salt and pepper to taste

Mix all the ingredients in a large bowl, careful not to mush the avocados too much. We want it chunky. Once properly mixed and rested, add the pits into the guacamole and even out the top with a spoon or spatula. Add ½ lime to the top layer so you cover most of the surface with the juice. (The pits and lime will keep it from browning prematurely[2].) Cover with plastic wrap and refrigerate until served. Please don't make it too early before it's served. We'd love to have it around 5pm.

[1] The recipe was deemed "decent", despite being "ridiculously specific", by top Peruvian chef Martin Morales in the *Guardian*.

[2] This is not entirely true. *See* page 169 for an explanation.

LUTEFISK

To anyone who didn't grow up in Norway or Sweden, *lutefisk* may not sound like a very appetizing prospect. Dried white fish, usually cod, is soaked for days, then steeped in a lye (caustic soda) solution, and then rinsed extensively before being boiled. The result, typically served with boiled potatoes and green peas, has an intense pong and a strange jelly-like consistency.

FOOD HOTSPOTS:
THE LITERARY HAUNT

Grand Café, Oslo – Every day at 2pm, the playwright Henrik Ibsen would stop by this Oslo institution (opened 1874) for lunch. Habitually, he would order an open sandwich, a beer and a schnapps – and often a *pjolter* (whisky and soda). The artist Edvard Munch, another regular, painted Ibsen sitting by the window with a newspaper. The pair are depicted together in a panoramic 1928 mural by Per Krohg, which can still be seen in the café today, although it's said they fell out when Ibsen once tried to pick up the artist's unpaid tab.

Peter Cat, Tokyo – Before he became a famous novelist, Haruki Murakami owned a cat-themed jazz club in the suburbs of Tokyo, near Kokubunji Station, later moving it to the central Sendagaya neighbourhood (a noodle joint named Jamaica Udon now occupies the spot). Peter Cat served coffee during the day, drinks and food at night, and hosted live jazz at the weekends. The second club closed in 1981. Murakami fans now congregate at Cafe Rokujigen in the Suginami neighbourhood to read his work and listen to some of his favourite music.

Musso & Frank, Los Angeles – As Hollywood began to lure literary authors in the 1930s with screenwriting dollars, this grill on Hollywood Boulevard (established 1919) became a refuge for the likes of F Scott Fitzgerald, William Faulkner and Raymond Chandler, who wrote chapters of *The Big Sleep* here. Other clientele over the years included Dorothy Parker, Charles Bukowski, Kurt Vonnegut – and the rather ubiquitous person discussed on page 107.

MORE BOOKISH HANGOUTS

Les Deux Magots, Paris – Simone de Beauvoir, André Gide, Arthur Rimbaud

Antico Caffè Greco, Rome – Lord Byron, John Keats, Henrik Ibsen and Hans Christian Andersen

White Horse Tavern, New York – Anaïs Nin, Jack Kerouac, Dylan Thomas

DINNER AT THE MOVIES

Food and cinema have a rich and enduring relationship and it's nearly impossible to narrow down the great restaurant scenes in the movies to just six. What about the diner scene in *Heat*? The deli scene in *When Harry Met Sally*? Every other scene in *The Cook, the Thief, His Wife & Her Lover*? The list could go on…

Louis' Restaurant in *The Godfather* (1972)

<u>What happens:</u> Michael Corleone (Al Pacino) shoots Virgil Sollozzo and Captain Mark McCluskey in an Italian restaurant in the Bronx, leading to the outbreak of the Five Families War in Francis Ford Coppola's mafia classic.

<u>On the table:</u> Italian table wine, antipasto salad, veal marsala (the veal is the best in the city, according to Sollozzo). The table is cleared at the end of the scene by McCluskey, who falls onto it after being shot in the head.

<u>Background:</u> The scene was filmed at an actual Bronx restaurant called Old Luna, with the real owner and his wife playing the hosts.

<u>Scores:</u> Food 9/10. Ambience 0/10

A French restaurant in *Monty Python's The Meaning of Life* (1983)

<u>What happens:</u> Mr Creosote (Terry Jones), an obese gourmand with a penchant for projectile vomiting, sits down to his final meal.

<u>On the table:</u> Moules marinière, pâté de foie gras, beluga caviar, eggs Benedictine, leek tart, frogs' legs amandine and oeuf de cailles Richard Shepherd on a bed of puréed mushroom – all mixed up in a bucket with fried eggs on top (a double helping). Jugged hare (very high, with a sauce which is very rich, with truffles, anchovies, Grand Marnier, bacon and cream). Six bottles of Château Latour '45 and a double Jeroboam of Champagne, six crates of brown ale. An entire pineapple. And, finally, a "waffer-thin" mint.

<u>Background:</u> The Pythons spent a whole week filming the sequence at Seymour Leisure Centre in Paddington, London. Within 12 hours of the explosion scene, which required thousands of gallons of minestrone "vomit" to be hurled at the walls, a couple used the room for their wedding, according to Michael Palin.

<u>Scores:</u> Food 7/10. Presentation 0/10

Sushi restaurant in *Oldboy* (2003)

What happens: Soon after his release from 15 years of imprisonment by a mysterious captor, Oh Dae-su (Choi Min-sik) walks into a sushi restaurant and tells the female chef behind the counter that he wants to eat a "living thing".

On the table: A live octopus. Oh Dae-su bites off its head and stuffs the rest into his mouth. As he chews, the tentacles writhe around his hand and face. Seemingly impressed by this display, the chef (who will play a key role in the film) puts her hand on his. Oh Dae-Su promptly – and not surprisingly, given what he's just eaten – blacks out.

Background: Live octopus is eaten in South Korea (the practice is known as *sannakji*) but it's usually sliced in advance. Choi, a Buddhist, had to eat four octopi, "with the tentacles really moving around in a good way". The scene was filmed at a restaurant called Gozen in the port city of Busan and Choi allegedly apologized to each octopus before biting its head off.

Scores: Freshness: 10/10. Appeal: 0/10

MORE CLASSIC RESTAURANT SCENES

North By Northwest (1959) – Cary Grant and Eva Marie Saint dine together on the 20th Century Limited train to Chicago, in one of the greatest seduction scenes in cinema. She's already eaten. He goes for the brook trout ("A little trouty," she tells him, "but quite good") and a Gibson cocktail.

Inglourious Basterds (2009) – Three years after killing her entire family, Christoph Waltz's SS colonel sits down in a Paris restaurant with Melanie Laurent's cinema projectionist to eat a very tense couple of apple strudels with cream.

Tampopo (1985) – While his superiors all order the same thing (sole meunière, consommé and a beer) at a fancy French restaurant, a young office worker goes to town on the menu – to the horror of his colleagues – ordering quenelle with caviar sauce, escargots wrapped in pastry and a 1981 Corton Charlemagne.

POETIC INTERLUDE

This is just to say

I have eaten
the plums
that were in
the icebox
and which
you were probably
saving
for breakfast
Forgive me
they were delicious
so sweet
and so cold

– William Carlos Williams (1934)

MANNA

When Moses led the Israelites out of Egypt and into the barren Sinai desert, God promised: "I will rain down bread from Heaven for you." The resulting substance, referred to in the Bible as manna, was "white like coriander seed and tasted like a wafer made with honey". It's impossible to identify this miraculous food, which sustained the Israelites over the next 40 years, though many have tried. One candidate is resin from the tamarisk tree. Another is a lichen, *Lecanora esculenta*, which can be dislodged from rocks during high winds and deposited on human settlements. A third is crystallized honeydew secreted by scale insects. It's unlikely that any of these could sustain a population for any serious length of time.

DON'T COOK THIS AT HOME!

It's easy to laugh at the culinary missteps of times past, particularly the 1950s and '60s when gelatin was in vogue and unholy flavour-marriages were being consummated at dinner parties everywhere. In full awareness that future generations will look back at our own food crimes and mock, here are four such ghastly creations.

Summer Salad Pie
Source: *Dinner in a Dish*, Betty Crocker, 1963
Offending combination: Lemon-flavoured gelatin mixed with tomato sauce, vinegar, celery, olives and onion. This is then poured into a cheese pie case, chilled and topped with tuna salad. Lemon-jelly-tuna-cheese!

Liver Sausage Pineapple
Source: *Better Homes and Gardens New Cookbook*, 1953
Offending combination: A gruesome mix of liver sausage, Worcestershire sauce, lemon juice and mayo is coated with gelatin, studded with olives and topped with a "real pineapple top" – because who can resist the thought of a liverwurst pineapple?

Pork Tenderloin with Bananas and Curry
Source: *The Nordic Cookbook*, Magnus Nilsson, 2015
Offending combination: It's all in the title. Fried banana halves are placed on top of the browned pork, then drenched in cream with curry powder. The whole thing is baked until the surface is golden. This is one dish from 1970s Scandinavia that Nilsson, the chef-owner at Fäviken, could have left untroubled in his titanic survey of Nordic cuisine (which also includes recipes for puffin and whale).

Crown Roast
Source: Unknown – reproduced by the *River Front Times* in 2009
Offending combination: Four cans of luncheon meat sliced in half, coated with orange marmalade and baked, then arranged in fortress-like formation on a platter with canned asparagus, pineapple and potatoes. God help us all.

THE SHAPES OF FOOD: BIRDS ON THE TABLE

The variety of birds that are commonly eaten by humans is vast. This list covers both domesticated and wild birds, ranking them by average size – for an 18-bird roast, perhaps? It ranges from the very tiny (adult ortolans weigh the same as an AA battery) to the huge (a fully-grown ostrich could balance two adult humans on a seesaw).

Ortolan

20–25g (¾–1oz)

A forbidden delicacy in France (due to its scarcity), where they are usually eaten in a single mouthful.

Japanese quail

90–100g (1¾–1½oz)

Domesticated in Japan a millennium ago, now farmed worldwide.

Snipe

80–140g (2¾–5oz)

An elusive marshland dweller, notoriously difficult to shoot, often hunted at night.

Woodcock

310g (11oz)

A Eurasian, prized forest-dweller. Usually roasted whole and served on toast.

Grey partridge

350–450g (12oz–1lb)

Rotund bird with a subtle flavour which is underappreciated.

Puffin

500g (1lb 2oz)

Lundi in Icelandic, puffin meat is usually smoked or grilled.

Wood pigeon

450–550g (1lb–1lb 4oz)

A larger, more voracious cousin of the urban pigeon.

Rock ptarmigan

430–740g (15oz–1lb 10oz)

Aka snow chicken, a key element in many Icelandic feasts.

Red grouse

600g (1lb 5oz)

Plump moorland bird with dark red meat and intense flavour.

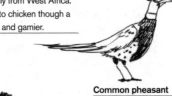

Guineafowl
1.3kg (3lb)
Originally from West Africa.
Similar to chicken though a
bit drier and gamier.

Tinamou
700g–1kg (1lb 9oz–2lb 4oz)
Common to South American
grassland, red-winged and
resembles small, squat
chickens.

Common pheasant
500g–3kg (1lb 2oz–6lb 8oz)
One of the most plentiful
European game birds, known
for its strut.

Chicken
2–5.5kg (4lb 8oz–12lb 2oz)
Broiler chickens, raised for
meat, are usually slaughtered
at 14 weeks (free-range).

Pekin duck
3–4kg (6lb 8oz–8lb 10oz)
Domesticated from mallard in
China, often used for Peking
duck.

Grey goose
Up to 10kg (22lb)
Kept domestically as poultry
since ancient times and a
traditional festive roast.

Ostrich
63–145kg (139–320lb)
Tastes like beef but with less
fat. Best feed-to-weight ratio
of any land animal.

Emu
18–60kg (40–132lb)
Native to Australia where its
red meat has been eaten for
thousands of years.

Turkey
Up to 39kg (86lb)
Large American domestic
bird forever linked with
Thanksgiving and Christmas.

GREAT MEALS IN LITERATURE: LUNCH

In the opening pages of **Anna Karenina** (1877) by Leo Tolstoy[1], Prince Stepan Arkadyevich Oblonsky and his friend Konstantin Levin go for lunch together at the Anglia. A clingy Tartar waiter ushers them into a private room and hands Oblonsky the menu, mentioning that they have fresh oysters. Perfectly at home in this rather showy Moscow restaurant, Oblonsky quizzes the waiter on the oysters' provenance, then turns to Levin, who is less enthralled by his surroundings.

> "In that case, shouldn't we begin with oysters, and then change the whole plan? Eh?"

> "It makes no difference to me. I like *shchi* [cabbage soup] and *kasha* [a thick gruel] best; but they won't have that here."

> "*Kasha à la Russe*, if you please?" the Tartar said, bending over Levin like a nanny over a child.

Levin has more important things on his mind than fancy food: he has just travelled from his country estate to propose marriage to Oblonsky's sister-in-law Kitty. Nevertheless, he gives his friend the go-ahead to order as he pleases, and Oblonsky turns back to the waiter, prompting a beautifully observed moment of social comedy.

> "Well, then, my good man, bring us two – no, make it three dozen oysters, vegetable soup…"

> "*Printanière*," the Tartar picked up. But Stepan Arkadyevich evidently did not want to give him the pleasure of naming the dishes in French.

> "Vegetable soup, you know? Then turbot with thick sauce, then… roast beef – but mind it's good. And why not capon – well, and some stewed fruit."

The Tartar, remembering Stepan Arkadyevich's manner of not naming dishes from the French menu, did not repeat them after him, but gave himself the pleasure of repeating the entire order from the menu: "*Soupe printanière, turbot sauce Beaumarchais, poularde à l'estragon, macédoine de fruits…*"

They order wine and five minutes later the waiter rushes in with the first course.

Stepan Arkadyevich crumpled the starched napkin, tucked it into his waistcoat, and resting his arms comfortably, applied himself to the oysters.

"Not bad," he said, peeling the sloshy oysters from their pearly shells with a little silver fork and swallowing them one after another. "Not bad," he repeated, raising his moist and shining eyes now to Levin, now to the Tartar.

Levin ate the oysters, though white bread and cheese would have been more to his liking.

Later in the meal, both men will get distracted from their food by a conversation about love, fidelity and the general unruliness of life. First, though, Levin reflects on the differences between country people like himself, who can't fathom the appeal of fine dining, and city slickers like Oblonsky who positively thrive on it.

"…it seems wild to me that while we countrymen try to eat our fill quickly, so that we can get on with what we have to do, you and I are trying our best not to get full for as long as possible, and for that we eat oysters…"

"Why, of course," Stepan Arkadyevich picked up. "But that's the aim of civilization: to make everything an enjoyment."

"Well, if that's its aim, I'd rather be wild."

[1] From Richard Pevear and Larisa Volokhonsky's translation (2000).

WHAT THE WORLD IS EATING

Seventy-five per cent of the world's food comes from just twelve plants and five animal species – which is not terrific news when it comes to the biodiversity of our planet. Here are the top five crops and types of livestock we produce on earth.

Crop	Production (million tons, 2013)
Sugar cane	1,877.1
Maize	1,016.7
Rice	745.7
Wheat	713.1
Potatoes	368.0

Livestock	Production (million tons, 2013)
Pigs	115.0
Poultry	108.6
Cattle	68.0
Sheep & goats	13.9

Source: Food & Agriculture Organization of the United Nations

THE DIRTY DOZEN & THE CLEAN FIFTEEN

These are the fruits and vegetables most and least at risk of contamination by pesticide use, according to the Environmental Working Group. The implication, for the former, is that you're better off buying organic.

The Dirty Dozen (*descending order of risk*) – Apples, celery, sweet bell peppers, peaches, strawberries, nectarines, grapes, spinach, lettuce, cucumbers, blueberries, potatoes.

The Clean Fifteen (*ascending order of risk*) – Onions, sweetcorn, pineapples, avocados, cabbages, peas, asparagus, mangoes, aubergines, kiwis, cantaloupes, sweet potatoes, grapefruit, watermelon, mushrooms.

THE OLD MAN AND THE BAR

Of all the literary figures to ever hang out in bars, few have been more prolific in their patronage than Ernest Hemingway. The long, long list of drinking and dining establishments that claim his custom can be traced through Spain, Cuba, Florida, Idaho and of course Paris, where he did a great deal of hanging out in the 1920s. Here are a few of his favourite haunts that are still open today.

The Ritz, Paris *(1920s–40s)* – He claimed to liberate the hotel from the Nazis in 1944. Reportedly ran up a tab for 51 dry Martinis at the bar. "Whenever I dream of afterlife in Heaven," he wrote, "the action always takes place at the Paris Ritz."

Brasserie Lipp, Paris *(1920s)* – In *A Moveable Feast*, he recounts going to this Left Bank stalwart (much cheaper back then than it is today) for a litre of beer and *pommes à l'huile* with sausage.

Casa Botín, Madrid (now Sobrino de Botín) *(1920s–50s)* – Near the end of *The Sun Also Rises*, characters dine on roast suckling pig here, narrator Jake Barnes referring to it as one of the best restaurants in the world.

Sloppy Joe's, Key West, Florida *(1930s)* – Owner Joe Russell allegedly supplied Hemingway with Scotch during Prohibition. In return, Hemingway suggested the bar's name. It now hosts an annual Ernest Hemingway lookalike contest.

La Floridita, Havana *(1930s–40s)* – Drank Daiquiris here with friends including Errol Flynn, Gary Cooper, John Wayne and Ava Gardner. A life-size bronze statue of Hemingway props up the bar.

Harry's Bar, Venice *(1949–50)* – Drank a lot here during his winter in Venice. According to founder Giuseppe Cipriani, "He filled more pages of his cheque book than those of a medium-length novel." Mentioned the bar in his novel *Across the River and into the Trees*.

THE FIVE STAGES OF BOILING WATER

Different teas need to be steeped at different temperatures to ensure good flavour. Long before thermometers, the Chinese devised a method of estimating temperature by paying close attention to water as it comes to the boil. The metaphors used for each stage make this no-tech approach even more appealing.

Shrimp Eyes (70°C/158°F) – Tiny bubbles appear at the bottom of the kettle. <u>Good for:</u> Very fine green teas such as Japanese Sencha and Gyokuro.

Crab Eyes (80°C/176°F) – Bubbles grow slightly and wisps of steam appear. <u>Good for:</u> Bolder green teas such as Pinhead Gunpowder, delicate oolongs such as Oriental Beauty, and white teas such as Silver Needle.

Fish Eyes (85°C/185°F) – Larger bubbles rise to the top. The pitch of the kettle may change. <u>Good for:</u> Most oolongs such as Dongding or Qilan.

String of Pearls (90–95°C/194–203°F) – You see a constant stream of bubbles.
<u>Good for:</u> Most black teas such as Dianhong and Keemun, Pu-erh teas and heartier oolongs.

Raging Torrent (100°C/212°F) – Properly boiling. <u>Good for:</u> Strong black teas such as Earl Grey, builder's tea, and fruit and herbal infusions.

INSATIABLE APPETITES

Viewed from most angles, gluttony is not a good thing. It is sinful, it is shameful, it is sad. Occasionally, though, one encounters an all-consuming appetite and can't help but feel impressed. How did they manage to eat all that?! As noted gourmand Elvis Presley used to say, when justifying the amount he put away: "The input has to be as great as the output." Brian Shaw (*see* page 162) would agree.

» Even by restaurant critic standards, **Jonathan Gold** of the *LA Times* has a grand appetite. Fellow reviewer Robert Sietsema recalled in *The New Yorker* a day they spent together in New York: "We started with porchetta sandwiches, then went to David Chang's new bakery for focaccia with kimchi, then we had salty-pistachio soft-serve ice cream, cookies and coffee milk. Then we went to a pizzeria famous for its artichoke slice, where we also had a Sicilian slice, and then we took the train to Flushing and visited a new Chinese food court and had half a dozen Chinese dishes there. Then we went to the *old* food court down the street, visited three more stalls, and had a bunch of things, including lamb noodles, and then Jonathan had to go to dinner somewhere. After dinner, he stopped by my apartment, and we went to Ten Downing, where we ate another three-course dinner."

» The actor **Marlon Brando** had difficulties with his weight throughout his life, veering between crash diets and gorging sprees. Early in his career, he was known to eat peanut butter by the jarful, boxes of cinnamon buns and huge breakfasts consisting of corn flakes, sausages, eggs, bananas and cream, and pancakes drenched in maple syrup. He would devour up to six hot dogs at a time in late-night feasts at Pink's in Hollywood (they named an all-beef hot dog after him in 2012). Defying attempts by work colleagues and loved ones to regulate his diet, he would break refrigerator locks at night, flee film sets with giant tubs of ice cream, and enlist friends to throw burger bags over the gates of his Mulholland Drive estate.

» Journalist and author **Hunter S Thompson** wasn't known for his restraint when it came to intoxicants. With food, it seems, his appetites weren't much less modest, particularly when it came to breakfast, which he described in his autobiography, *The Great Shark Hunt*, as "a personal ritual that can only be properly observed alone, and in a spirit of genuine excess. The food factor should always be massive: four Bloody Marys, two grapefruits, a pot of coffee, Rangoon crepes, a half-pound of either sausage, bacon, or corned beef hash with diced chillies, a Spanish omelette or eggs Benedict, a quart of milk, a chopped lemon for random seasoning, and something like a slice of Key lime pie, two Margaritas, and six lines of the best cocaine for dessert... All of which," he concluded, "should be dealt with outside, in the warmth of a hot sun, and preferably stone naked."

MICHELANGELO'S SHOPPING LIST

In 1518, while Michelangelo was reconstructing the façade of the Basilica of San Lorenzo in Florence, he jotted down a quick list of groceries for an assistant to collect from the local market. Alongside these items – fish, bread, wine and so on – he made hurried sketches, not to relieve an irrepressible urge to draw but, probably, because the servant running the errand was illiterate.

The list reads: "*Pani dua* (two bread rolls), *un bochal di vino* (a jug of wine), *una aringa* (a herring), *tortegli* (stuffed pasta), *una insalata* (a salad), *quatro pani* (four bread rolls), *un bochal di tondo* (a jug of full-bodied wine), *un quartuccio di bruscho* (a quarter of dry wine), *un piatello di spinaci* (a dish of spinach), *quatro alice* (four anchovies), *tortelli* (stuffed pasta), *sei pani* (six bread rolls), *dua minestre di finochio* (two dishes of fennel), *una aringa* (again), *un bochal di tondo* (again)."

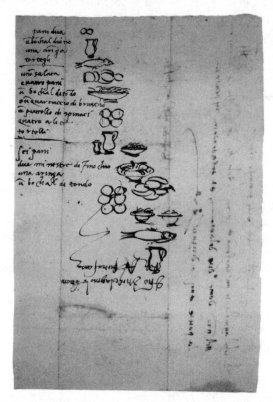

KNOW YOUR FOOD EMOJI

Emojis don't appear out of thin air: companies and private citizens petition for new inclusions and an industry body – the Unicode Consortium – decides, based on expected levels of usage and other factors, whether they are worth adding to the universal lexicon of cute pictograms. There are now 102 food and drink emojis in circulation out of a total of 2,623. Here are eight of them:

1. Paella: Outraged Spaniards protested against an inaccurate depiction of the Valencian dish in 2015, prompting a correction by Emojipedia a few months later, minus the offending prawns.

2. Aubergine: Instagram banned this unintentionally phallic emoji from its feeds in 2015, but took no similar steps against the banana.

3. Hot dog: Chicago hot-dog vendor Superdawg campaigned for the inclusion of the fast-food classic in 2015.

4. Pepperoni slice: Domino's Pizza allowed customers to text and tweet orders using the pizza emoji in 2015.

5. Taco: US chain Taco Bell secured nearly 33,000 signatures on a virtual petition in 2015 to create a taco emoji.

6. Peach: Apple redesigned their version of this emoji in 2016 to make it look less like a bum.

7. Naruto: Emojis originated in Japan, hence the bias towards foods that are obscure to many non-Japanese. This is a moulded and steamed fish cake often found in ramen.

8. Hot beverage: The first food or drink emoji, created in 2003.

FOOD AT THE EDGE OF THE WORLD

Great explorers of centuries past encountered numerous challenges on their voyages, including extreme weather, pestilence and piracy, but some of the most desperate hardships were food related. Many of the following pioneers weighed the prospect of glorious discovery against the risk of food poisoning, scurvy and starvation, and decided it was worth the gamble. Some had it significantly worse than others...

Amelia Earhart's first transatlantic flight (1928)
Oranges, egg sandwiches, pemmican (a meat paste), Drake's oatmeal cookies, Horlick's malted milk tablets, coffee, mineral water, Hershey's chocolate.

Zheng He's sea voyages from Asia to East Africa (1405–33)
As well as rice and other preserved foods, the Chinese explorer's enormous wooden ships carried large tubs of soil for growing fruit and vegetables.

Sir Edmund Hillary and Tenzing Norgay's ascent of Everest (1953)
Before their final climb to the top, the pair ate chicken noodle soup, sardines on biscuits, tinned apricots, dates, jam and honey, and hot water with lemon.

Hawai'iloa's discovery of Hawaii (around 5th century)
According to legend, the Polynesian explorer and his crew survived on taro root for the 2,400-mile journey.

Captain Robert Scott's race to the South Pole (1911–12)
In the hut at Cape Evans, Scott's party ate well: bread, rhubarb pie, curried and fried seal meat, turtle soup, stewed penguin breast. But on their final, ill-fated journey they were reduced to eating hoosh, a stew of pemmican and biscuits, bulked out with meat from their ponies.

Robert O'Hara Burke's crossing of Australia (1860–1)
Facing starvation on their fateful return journey to Melbourne, the
Irish-born explorer and his men were forced to eat their camels and
a horse, as well as purslane and a python. Aborigines at Cooper Creek
provided fish and beans. Burke's final meal was said to be nardoo
(a cake made from fern seed) and a crow.

Ferdinand Magellan's crossing of the Pacific (16th century)
Underestimating the vastness of the ocean, the Portuguese explorer and
his crew were reduced to eating worm-infested biscuits stinking of rat's
urine. Then they cooked and ate leather strips that had been softened
in sea water. Finally, before reaching land, they ate sawdust and rats.

Sir John Franklin's Arctic expeditions (early 19th century)
On an early voyage along the northern coast of Canada, the British
explorer's crew were reduced to scraping lichen off rocks and eating
shoes. Heading for the Northwest Passage in 1845, Franklin was much
better provisioned but the tinned food was contaminated and the crew
succumbed to lead poisoning, among other horrors. None survived.
His two ships weren't found until 2014 and 2016.

NIXTAMALIZATION

A complex process for preparing maize developed
by Aztecs and Mayans in Mesoamerica as early as
1500BC. It involves soaking and cooking the maize
in an alkaline solution, then washing and hulling
the grains. The end product is easier to grind than
unprocessed maize, the flavour is better – and the
nutritional value is so much higher that some
historians believe nixtamalization helped fuel the
rise of Mesoamerican civilization. The process
persists today in the manufacture of tortillas and
tamales, among many other maize-based foods.

TEN DEGREES OF PRESERVATION

Few foods last long without someone taking active steps to preserve them. Over millennia, humans have become very adept at protecting their larders from the enzymes and bacteria that cause food spoilage – and by encouraging other bacteria that have the opposite effect. As well as keeping food from going off, many of these methods (such as salting and smoking) can do wonders for flavour and texture.

Drying – Moisture is drawn out by heat so that micro-organisms cannot grow. (*Dried mint, sun-dried tomatoes, dried pasta*)

Salting – Moisture is drawn out by salt, suspending the action of enzymes and micro-organisms. (*Beef jerky, salt cod*)

Pickling – Acid conditions are created by adding a strong vinegar solution, or by encouraging lactic acid-producing bacteria with salt. This suspends the action of enzymes and other harmful bacteria. (*Gherkins, pickled herring, sauerkraut, kimchi*)

Canning – Oxygen is drawn out of a food container by heating it, creating a vacuum and killing bacteria. (*Tinned tuna, baked beans*)

Chilling & freezing – Spoilage is slowed – or almost completely suspended – by low temperatures. (*Refrigerated milk, frozen peas*)

Cooking – Proteins are denatured by the application of heat. Most bacteria and all moulds are killed. (*Boiled potatoes, roast chicken*)

Sugaring – Moisture is drawn from microbes, keeping food safe from spoilage. (*Fruit in syrup, jam*)

Smoking – Exposure to smoke dries food and seals the surface with an antiseptic coating. (*Smoked salmon, bacon*)

Jellying – A solid gel is formed through cooking, sealing out bacteria. (*Jellied eels, pâtés*)

Burial – Oxygen, moisture and heat are reduced by covering food in dry soil to prevent spoiling. (*Century eggs, fermented shark*)

SERIOUSLY NICHE COOKBOOKS

If you can imagine it, chances are it already exists and has been discussed in detail on the Internet. Has anyone ever written a book of reptile recipes, for example, or a guide to cooking under the influence? Google it and, yes, there it is (*The Iguana Cookbook* by George Cera; *The Drunken Cookbook* by Milton Crawford). Whatever wacky urge possesses you in the kitchen, there is probably a book out there to help you satisfy it.

If you want to...	Read this
Fry fish under the hood of your car	*Manifold Destiny: The One! The Only! Guide to Cooking on Your Car Engine!* by Chris Maynard & Bill Scheller
Eat the thing you've just run over with your car	*The Original Road Kill Cookbook* by BR "Buck" Peterson
Recreate your favourite airline food	*The Famous Airline Cookbook: A History and a Cookbook* by Jerry Honeywell
Make miniature cakes while staring at hunks in aprons	*Stud Muffins: Luscious, Delectable, Yummy (and Good Muffin Recipes, Too!)* by Judi Guizado, Shari Hartz & Gilda Jimenez
Prepare food with no clothes on	*Cooking in the Nude* by Stephen & Debbie Cornwell
Self-cater with a microwave	*Microwave Cooking for One* by Marie T Smith
Seek culinary advice from a 1990s gangster rapper	*Cookin' with Coolio* by Coolio
Eat the contents of your local aquarium	*Vancouver Aquarium Seafood Recipes* compiled by Ainley Jackson
Make dinner for (not out of) your beloved pet	*The Kitty-Cat Cookbook* by Barbara Ellen Benson

THE WORLD HOT DOG EATING RECORD

72 – The number of hot dogs and buns Joey "Jaws" Chestnut ate in ten minutes at the 2017 Nathan's Hot Dog Eating Contest in Coney Island, New York, to set a new competition record and take the mustard-coloured belt for the tenth time. Chestnut, of Vallejo, California, broke onto the competitive eating scene in 2005 when he won a deep-fried asparagus eating championship, consuming 6.3 pounds (2.8 kilograms) of asparagus in 11 ½ minutes (*see* page 69).

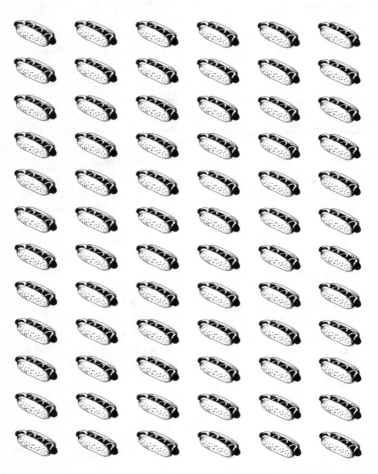

THE WOMAN WHO REVOLUTIONIZED THE KITCHEN

"I had never concerned myself with cooking in my life," **Margarete Schütte-Lihotzky** once said. And yet, her influence on the modern domestic kitchen was profound, even if she went unrecognized for most of her lifetime. Born in 1897, Lihotzky was the first female architecture graduate of the Vienna School of Arts and Crafts. Her interest in creating functional and sympathetic designs for social housing brought her to Germany in 1925 to work on the New Frankfurt project under architect Ernst May.

There, she created what became known as the Frankfurt Kitchen. It was a compact unit, measuring just 1.9 × 3.4m (6ft 2in × 11ft), but Lihotzky put great thought into how its user – invariably a woman – would move around the space as she worked. The attention to detail was unprecedented: drawer fronts were painted blue because research showed that flies avoided blue surfaces. Ten thousand of these kitchens were installed in the late 1920s, though the project petered out and Lihotzky's career suffered in the worsening political climate (she was imprisoned by the Nazis during the war). Few of her kitchens survive today. However, Lihotzky's project to liberate women from the drudgery of housework through efficient design – and the fact that she created the first fitted kitchen – had found recognition before she died in 2000 at the age of 102.

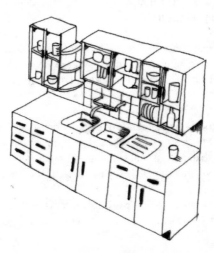

FOOD AVERSIONS OF THE FAMOUS

Few people love all kinds of food without exception, eating broccoli, tripe or fermented fish as readily as ice cream. Some aversions are innate (babies tend to be wary of the bitter tastes common in plant toxins) though most can be overcome by repeated exposure. Here's an array of well-known people resolute in their dislike of certain foods (for a comprehensive glossary of food phobias, *see* page 136).

Samuel Johnson disliked **cucumbers**. According to his biographer James Boswell, he felt "a cucumber should be well sliced, and dressed with pepper and vinegar, and then thrown out as good for nothing".

Jean-Paul Sartre abhorred **crabs and lobsters**, which reminded him of insects.

Samuel Beckett had similar misgivings about crustaceans. Dining at Aux Iles Marquises in Paris, he asked to be seated far away from the **lobster and trout** tank because it upset him.

Coco Chanel steered clear of **onions**. "I don't like food that talks back to you after you've eaten," she said.

Salvador Dalí detested **spinach** "because of its utterly amorphous character," he said, adding: "The very opposite of spinach is armour. That is why I like to eat armour so much, and especially the small varieties, namely, all shellfish."

George HW Bush banned **broccoli** from the White House: "I haven't liked it since I was a little kid and my mother made me eat it," he said. "And I'm president of the United States and I'm not going to eat any more broccoli."

Alfred Hitchcock was frightened of **eggs**. "Worse than frightened, they revolt me," he told Oriana Fallaci. "Have you ever seen anything more revolting than an egg yolk breaking and spilling its yellow liquid? Blood is jolly, red. But egg yolk is yellow, revolting. I've never tasted it."

Sigmund Freud refused to eat **cauliflower** and **chicken**. "One should not kill chickens," he said. "Let them stay alive and lay eggs."

Judy Garland avoided **breakfast cereal** and **orange juice**, which she said gave her hives.

Elvis Presley enjoyed many types of food but not **seafood**.

Pliny the Elder had no time for **artichokes**, writing them off as one of "the earth's monstrosities".

Jeffrey Steingarten, food critic for *Vogue,* confessed to a swathe of food aversions in his 1999 book, *The Man Who Ate Everything.* These included **anchovies** ("Why would anybody consciously choose to eat a tiny, oil-soaked, leathery maroon strip of rank and briny flesh?") and **lard** ("The very word causes my throat to constrict and beads of sweat to appear on my forehead").

Ayn Rand despised **salads**.

CERTAIN FOODS ARE MORE LIKELY TO CAUSE OFFENCE

Beetroot didn't appeal to Albert Einstein. Nor does it to Michelle or Barack Obama (who also avoids asparagus, mayonnaise, and salt and vinegar crisps).

Coriander was disliked by Julia Child and finds no favour with Yo-Yo Ma, Ina Garten or chefs Angela Hartnett and Fergus Henderson, the latter describing it to me as a "bully".

Caviar fails to impress actors Naomi Watts and Jennifer Aniston. "It's just a shitty taste," Aniston said of the delicacy in an interview.

HOW LONG DOES IT TAKE TO DIGEST A DUCK?

In *The New Century Cookery Book*, first published in 1901, Swiss-born cookery teacher Charles Herman Senn printed an intriguing table showing "the time required for the complete digestion of the most common articles of food" from apples to venison. I'm reproducing a slightly abridged version here. "It should be remembered," Senn adds, "that the addition of fatty substances or acidity to food makes digestion more slow, while the addition of seasonings will further it."

Food	Time	Food	Time
Apples, sweet	**1h30m**	Lobster (boiled)	**4h00m**
Artichokes (boiled)	**1h05m**	Mackerel (grilled)	**3h45m**
Asparagus (boiled)	**1h30m**	Milk (raw)	**2h15m**
Bananas (baked)	**1h30m**	Mushrooms (boiled)	**5h00m**
Beans (boiled)	**3h45m**	Nuts	**4h00m**
Beefsteak (grilled)	**3h00m**	Onions (boiled)	**3h30m**
Beets (boiled)	**3h45m**	Orange	**2h45m**
Bread (baked, fresh)	**3h30m**	Oysters (raw)	**2h55m**
Butter (melted)	**3h30m**	Peas, green (boiled)	**2h35m**
Cabbage (boiled)	**4h30m**	Pork, fat (roasted)	**5h15m**
Cauliflower (boiled)	**2h00m**	Potatoes (boiled)	**3h30m**
Cheese (melted)	**3h30m**	Potatoes (fried/baked)	**2h30m**
Chicken (boiled)	**2h00m**	Rabbit (stewed)	**4h30m**
Cod (boiled)	**2h00m**	Rice (boiled)	**1h00m**
Cucumber (raw)	**4h45m**	Salmon (boiled)	**2h00m**
Duck (roasted)	**4h00m**	Sausage (grilled)	**3h20m**
Eggs (hard-boiled)	**4h00m**	Soup, bean (boiled)	**3h00m**
Eggs (soft-boiled)	**3h00m**	Spinach (boiled)	**1h30m**

FERGUS HENDERSON'S DEEP-FRIED SQUIRREL

I found this recipe on a deckchair in the South London garden of Fergus and Margot Henderson while I was interviewing them for *The Gannet* – it was printed on the deckchair fabric. Fergus is the founder of St John restaurant and a master of cooking with unusual meats. If there's anyone I'd trust to instruct me on how to deep-fry a squirrel, it's him. I took a photograph of the recipe and am reprinting it here with Fergus's permission.

One squirrel per person

Pig's trotters, 1 to every 4 squirrels
Whole heads of garlic,
1 to every 4 squirrels
Light chicken stock
A glass of Vielle Prune
Squirrel, cut into 2 shoulders,
saddle, and 2 hind legs
Seasoned flour
Eggs whisked with a fork into
egg wash
Very fine breadcrumbs (the fineness of the crumbs is vital, so you end up with just a whisper of crispiness between you and the squirrel)
Vegetable oil, for frying

Place trotters, garlic and stock (enough to eventually cover your jointed squirrel) in a pot. Bring to the boil, then reduce to a simmer for 3 hours, allowing the trotters to emit their goodness.

Take off the heat, add glass of Vielle Prune and then the squirrel pieces. Cover pot and put in a very gentle oven for 1.5 to 2 hours, checking with a knife how tender the flesh feels. When satisfied, remove pot from oven and squirrel from pot to cool down. Keep cooking liquor, as it will make an excellent and most reviving squirrel broth.

Once the squirrel is cold, lightly dust with flour, coat with the egg wash and cover with breadcrumbs. Heat the oil in a pan and deep-fry your squirrel until golden.

A WHISTLESTOP TOUR OF RAMEN

The great Japanese noodle soup known to the (increasingly addicted) world as ramen may appear simple enough on the surface, but its history and regional variations are as tangled as the noodles lurking within. At its most basic, ramen consists of a broth made from slow-boiling meat bones (or sometimes fish, or vegetables); this goes in a bowl with wheat noodles, a salty seasoning and various goodies on top. But, as you'll see from this super-compressed guide, there are endless different ways to construct your ramen.

KEY VARIABLES

Heaviness – Is your ramen *kotteri* (rich) or *assari* (light)? This will depend mainly on the contents of your broth and how long it's been boiling.

Broth base – The best-known broth is probably *tonkotsu*, made by boiling pork bones for many hours until rich, sticky and opaque. Other bases include chicken, beef and fish bones – or sea kelp or dried seafood for lighter broths. Additional elements include *kombu*, garlic, ginger, spring onions, leeks and mushrooms.

Seasoning – Saltiness is likely to come from *shio* (sea salt), *shoyu* (soy sauce) or *miso* (fermented bean paste). It can be added to the bowl or mixed straight into the soup base.

Noodles – Fresh, dried or instant are the main variables, but degrees of thickness, firmness, curliness and bounciness are also significant. The basic ingredients are wheat flour, salt, water and *kansui*, an alkaline water that gives the noodles their golden hue.

Toppings – The obvious one is pork – roast, shredded or ground – though vegetables or seafood (crab, scallops, mussels and so on) sometimes take the starring role. Slow-cooked eggs are an important component. Also: mushrooms, spring onions, pickled ginger, kimchi, nori, butter – and the various condiments (*togarashi*, sansho pepper, for example) that you add at the table.

FIVE SIGNIFICANT JAPANESE RAMEN REGIONS

Sapporo – The birthplace of *miso* ramen (fermented bean paste is the main seasoning), this often snowy northern city is known for its rich, fatty soups. Toppings include roast and ground pork, spring onions, beansprouts, cabbage, ginger and garlic. A pat of cold butter is often stirred in at the end.

Kitakata – A small town in Fukushima province, north of Tokyo, with a high density of ramen shops. Light, clean soups are seasoned with *shoyu* and served with flat, wide and curly noodles. Minimal toppings include roast pork, spring onions and bamboo shoots.

Tokyo – Although the capital hosts a crazy variety of styles, there is a traditional (and very popular) Tokyo ramen made with pork and chicken broth and seasoned with *shoyu*. Dashi broth made from smoked bonito flakes and sea kelp is often added. Toppings include roast pork, spring onions, nori and bamboo shoots.

Yokohama – When Chinese traders brought ramen to Japan in the late 19th century, Yokohama, just south of Tokyo, was by most accounts, where it first took hold. The speciality today is known as *ie-kei* ramen, a salty, fatty *tonkotsu*–style soup with *shoyu* seasoning. Toppings include roast pork, nori, stewed spinach and a boiled egg.

Hakata – A district of Fukuoka city in the south, Hakata is renowned for its creamy pork-bone *tonkotsu* broth (which originated in nearby Kurume), eaten with roast pork and thin, almost raw noodles. Self-service toppings include garlic, pickled ginger, sesame seeds and spicy pickled mustard greens.

FAMOUS TALENTS IN THE KITCHEN

Midway through his insanely long, wildly digressive and highly entertaining 2017 book *Einstein's Beets*, the American writer Alexander Theroux (brother of Paul) hits us with a diverse list of famous people who knew a thing or two in the kitchen. They range from Sheryl Crow to Jean Cocteau to the Biblical prophet Ezekiel – a motley crew. I've borrowed from Theroux's list and added a few names of my own, below.

Jackson Pollock
Style: Homely American. The abstract expressionist was a keen gardener and baker.
Typical dish: Pollock's apple pie routinely won top prize at a local fair in East Hampton, NY.

Sylvia Plath
Style: The poet and novelist was a keen baker. "Instead of … writing – I go make an apple pie, or study the *Joy of Cooking*, reading it like a rare novel," she wrote in 1957.
Typical dish: Tomato-soup cake with cream cheese frosting (!)

Boy George
Style: The pop star discovered macrobiotic cooking in the 1990s and went on to publish *Karma Cookbook: Great Tasting Dishes to Nourish Your Body and Feed Your Soul* in 2001.
Typical dish: Watercress and shiitake salad.

Georgia O'Keeffe
Style: Simple, healthy food using local, organic produce – the US artist, who died in 1986, was ahead of her time.
Typical dish: Watercress soup.

John Steinbeck
Style: The novelist was a true locavore: fishing and clamming on Long Island, foraging for greens in England, even buying a cow in California so he could have his own butter and cheese.
Typical dish: Pork pozole – a Mexican stew made with hominy (dried maize), which he cooked while travelling through the badlands of North Dakota.

Frida Kahlo

Style: The artist learned to cook after marrying Diego Rivera in 1929 and would turn out Mexican dishes which, as described in the book *Frida's Fiestas*, were almost as vibrant and colourful as her paintings.

Typical dishes: Chillies in walnut sauce, tamales with chicken picadillo, chilaquiles in green sauce, squash blossom soup, pork with nopales.

Scarlett Johansson

Style: "I cook a lot of simple things, just dinner at home," she told *Saveur* magazine, adding: "I like to bake cakes and pastries and cookies and things like that. And pie."

Typical dish: Turkey Bolognese. "I like undercooked pasta," she said. "Really undercooked. Very al dente."

Louis XV

Style: The 18th-century French monarch was an enthusiastic cook and would lecture his chefs on the best ways to produce something as simple as an omelette.

Typical dish: Onion soup. Legend has it that he made the first one.

OMAKASE

It comes from the Japanese word for "entrust" and is used to describe a meal where the chef decides what the diner eats. It may include sushi, tempura, teriyaki and vegetable dishes, but the star of the show – and the factor that makes *omakase* so hard on the wallet – is ultra-fresh seafood, often flown in at great expense.

DESERTS OF ASIA

i. Baklava (Turkey) – Popular throughout the Middle East, baklava is layered filo pastry with chopped nuts (usually hazelnuts or pistachios) soaked in copious amounts of honey or syrup.

ii. Faloodeh (Iran) – A cool treat in the hot summer months, this Shiraz speciality consists of semi-frozen vermicelli noodles mixed with rosewater syrup. It's often served with lime juice and pistachios.

iii. Chakchak (Uzbekistan) – Short sticks of dough are deep-fried and mixed with a honeyed syrup, then piled into mounds. Versions of this are made throughout the former Soviet Union.

iv. Sheer payra (Afghanistan) – Milk fudge studded with pistachios and flavoured with rosewater and cardamom. Often served with tea or coffee.

v. Gulab jamun (India) – Milk is slowly reduced to form a rich curd, then shaped into spheres the size of golf balls. These are soaked in a syrup infused with cardamom and rosewater.

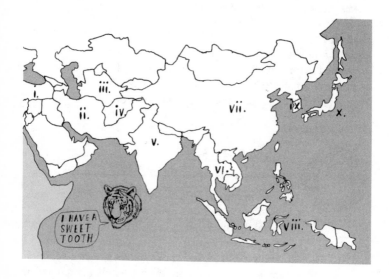

vi. Khao neow ma muang (Thailand) – Chilled slices of mango, warm sticky rice, coconut cream syrup over the top: unsurprisingly one of Thailand's most popular sweet treats.

vii. Dòu shā bāo (China) – Red (or adzuki) beans are a key element in many East Asian desserts. In China, sweetened bean paste can often be found inside a steamed bāo bun.

viii. Gethuk (Indonesia) – A cassava dessert? Why not. For this Javanese confection, the root is boiled, mashed and mixed with grated coconut, sugar and sometimes vanilla.

ix. Gyeongdan (Korea) – Glutinous rice flour is boiled in water, rolled into little balls and sweetened with honey or syrup. Some are stuffed, others are coated with green tea powder or black sesame seeds.

x. Tokoroten (Japan) – Trust Japan to acquire a taste for a jellified seaweed dessert. *Tengusa* or *ogonori* (types of red seaweed) are boiled, pressed and turned into clear noodles, which can be sweetened with sugar or syrup.

DINNER WITH JOAN D

In 2014, to drum up support for a documentary about the American author and journalist Joan Didion on Kickstarter, the producers offered to reward pledges of $50 or more with a PDF of "Joan's personal recipes for pork roast with corn soufflé, gumbo, linguini Bolognese… all perfectly preserved in her handwriting". It also included a recipe for a parsley salad with a simple oil and vinegar dressing and a grating of Parmesan. It stands out in part because it gives us a sense of how popular Didion's dinner parties must have been back in the day. Its serving quantity? 35 to 40 guests.

THE SHAPES OF FOOD: EDIBLE FLOWERS

They're all the rage lately – strewn over salads, adding dazzle to desserts, prettifying dishes for Instagram – but edible flowers have been used in cooking for millennia, finding favour in Ancient Greece, Rome and China. Before we proceed, be advised that many flowers are not edible, and even those that are should be treated with caution (avoid flowers from polluted areas, eat only the petals, try a small amount first in case you're allergic).

Bee balm

Monarda didyma

Traditionally used by Native Americans to make a medicinal tea. Also good for seasoning meats.

Red-pink • Peppermint

Borage

Borago officinalis

Used in salads, stews and as a base for soup in France and Italy. Features in Pimm's No 1.

Azure blue • Cucumber

Chamomile

Anthemis nobilis

Much consumed as tea in Victorian times. Said to be good for the digestion.

White/orange • Grassy

Chrysanthemum

C. morifolium etc

The petals of different varieties are used in salads, tea and wine.

Various • Various

Dandelion

Taraxacum officinale

Flower buds are often pickled, the heads used to make wine.

Yellow • Bittersweet

Hibiscus

Hibiscus sabdariffa etc

Makes a delicious cordial or tea. Also good in desserts.

Various • Cranberry

Hollyhock

Hollyhock

Alcea rosea

Soothing for sinus and
stomach complaints, turns
up in teas, salads, stews.

White to red • Mild

Jasmine

Jasmine

Jasminum sambac

Used for scenting tea in
China. (All other varieties are
poisonous.)

Yellow-white • Fragrant

Marigold

Marigold

Calendula officinalis etc

Not all varieties are edible.
Eat raw in salads; scatter
over an omelette or stir-fry.

Orange/yellow • Tangy

Squash Blossom

Squash blossom

Cucurbita pepo etc

Stuff with herbs and cheese
and deep-fry – as the Italians
do.

Orange/green • Squash

Nasturtium

Nasturtium

Tropaeolum majus

Excellent in salads, sprinkled
over pasta dishes or stirred
into omelettes.

Yellow/red • Peppery

Rose

Rose

Rosa alba etc

Linked with sweet dishes:
sorbets, jellies, jams. Darker
petals have more flavour.

Various • Various

Sunflower

Sunflower

Helianthus annuus

Add petals and toasted
seeds to salads, toss
blanched buds in garlic
butter.

Yellow • Nutty

Sweet Cicely

Sweet cicely

Myrrhis odorata

Long used as a sugar
substitute. Good in drinks,
fruit salads.

White • Aniseed

Violet

Violet

Viola odorata

Made into wine in Persia and
Rome, often candied in the
Tudor era.

Violet • Sweet

A MENU IN SONG TITLES

Good Evening (Mac Miller). *Table For Two* (Loretta Lynn). *Not A Problem* (Black Lips). *Come On In* (Eminem). *Follow Me* (John Denver). *May I Take Your Order?* (2 Chainz).

To Drink
Gin and Juice (Snoop Dogg)
Red, Red Wine (UB40)

Starters
Green Onions (Booker T & the MGs)
Eggs and Sausage (Tom Waits)
Chicken Grease (D'Angelo)

Main course
Fish, Chips and Sweat (Funkadelic)
Pork and Beans (Weezer)
Chop Suey! (System of a Down)
Jambalaya (Hank Williams)

Sides
Vegetables (The Beach Boys)
(Do the) Mashed Potatoes (James Brown)
Truffle Butter (Nicki Minaj)

Dessert
Wild Honey Pie (The Beatles)
Birthday Cake (Rihanna)
Ice Cream for Crow (Captain Beefheart)
Apple Peaches Pumpkin Pie (Jay and the Techniques)
Lemon Incest (Serge and Charlotte Gainsbourg)

Digestif
Whiskey in the Jar (Thin Lizzy)
Black Coffee in Bed (Squeeze)

THE FOOD OF DREAMS

In his 2017 book *Everybody Lies: What the Internet Can Tell Us About Who We Really Are*, American data scientist Seth Stephens-Davidowitz sets out to discover whether – as one might expect from reading the work of Sigmund Freud – phallic foods such as bananas and cucumbers appear more frequently in our dreams than their non-phallic counterparts. He did so by analyzing a huge trove of data from *Shadow*, an app which asks users to record their dreams, and from this he was able to estimate which foods arose most frequently. The results, which Stephens-Davidowitz kindly shared with me, may have left the Viennese psychoanalyst feeling a little deflated.

1. Chocolate	11. Popcorn
2. Pizza	12. Rice
3. Chicken	13. Apple
4. Bread	14. Salmon
5. Sandwich	15. Lobster
6. Hamburger	16. Potato
7. Honey	17. Squash
8. Cheese	18. Sushi
9. Banana	19. Ketchup
10. Mushroom	20. Spaghetti

PERCEBES

Also known as the goose barnacle, this is in fact a crustacean which attaches itself to rocks on the water's edge. It's weird-looking, with a scaly tube mounted on bony pads that resemble a hoof, but it is prized as a delicacy in Portugal and the Atlantic coast of Spain. So great is its appeal that *percebeiras*, or barnacle gatherers, risk their lives to harvest it from wave-battered rocks and cliff bases. Usually lightly boiled but sometimes eaten raw – the inner tube pulled from the scaly outer case.

WHERE DID YOUR BURGER COME FROM?

It's the most everyday of meals: burger and fries, followed by ice cream and chocolate sauce, all washed down with a glass of cola on the rocks. What could be more straightforward? But take a closer look and a tangle of stories stretching back thousands of years to the earliest days of agriculture emerges. Let's (at least begin to) break it down.

BURGER AND FRIES

» **Beef** (patty) and **cheese**: All modern domesticated cows are believed to be descended from a single herd of wild oxen, which roamed around Europe and Asia 10,500 years ago.

» **Wheat** (bun): First cultivated in the Middle East around 9600BC, this grass now covers about 2,253,290 square kilometres (870,000 square miles) of the earth's surface.

» **Tomato**: Native to western South America and Central America.

» **Lettuce**: First cultivated in Ancient Egypt.

» **Potato** (fries): Domesticated in what is now southern Peru and northwest Bolivia between 8000 and 5000BC.

The origin of the **hamburger** is a contested issue: early versions may have emerged in Hamburg, Germany, but the burger really took shape in America around the turn of the 20th century. Often cited as its inventor is Louis Lassen, who served ground meat between slices of bread at his Connecticut restaurant Louis' Lunch in 1900.

Fries arrived earlier than the burger: origin stories in Belgium and France go back as far as the 1600s.

ICE CREAM & CHOCOLATE SAUCE

» **Vanilla** (ice cream): The Totonac people of eastern Mexico are believed to have been the first to cultivate this orchid fruit.

» **Chocolate** (sauce): Indigenous to the region between southern Mexico and the northern Amazon basin, with evidence of cultivation in Mexico as early as 1750BC. Originally consumed as a drink, it was brought back to Europe by the Spanish conquistadores (as was vanilla).

You can go back a long way looking for the earliest iced desserts – ancient Greeks were eating snow mixed with fruit and honey in the 5th century BC – but it seems that **ice cream** as we know it today, made with milk or cream, took shape in 17th-century Italy and France.

← ANCIENT
BRAIN FREEZE

COLA

» **Cinnamon**: The dried bark of the *Cinnamomum zeylanicum* tree, indigenous to Sri Lanka, was introduced to Egypt as early as 2000BC and was greatly prized in Ancient Greece and Rome, though its true origin was a mystery in Europe until medieval times.

Cola was invented in 1886 by American pharmacist John Pemberton. Originally intended as a medicine, his Coca-Cola drink likely contained kola nuts (a source of caffeine) and extract of coca leaves. Pemberton died in 1888 and rights to his formula were bought by Asa Candler, who built the Coca-Cola company into a soft drinks giant. The recipe is famously a trade secret, but today it contains neither coca nor kola. Besides cinnamon, ingredients are believed to include vanilla, caramel colouring, a sweetener, phosphoric acid and some mixture of citrus oils.

UNLIKELY COOKBOOK AUTHORS

Vanity project. Cynical cash-in. Labour of love. Extended practical joke. There are many possible reasons for celebrities to write (or at least put their name on) a cookbook – it's often a dash of one and a sprinkling of the other. Mostly these books are to be avoided, but occasionally one comes along that merits further attention…

Andy Warhol – In 1959, when Warhol was still a little-known illustrator, he teamed up with the New York interior decorator Suzie Frankfurt to spoof the fashionable French cookbooks of the era. They produced 34 handmade copies of *Wild Raspberries*, in which Frankfurt's zany recipes (Roast Iguana Andalusian, Omelet Greta Garbo) were matched by Warhol's nicely camp illustrations. The book was forgotten until Frankfurt's son, Jaime, discovered a copy in his mother's papers and published it in 1997.

Liberace – The legendary pianist's love of food was such that he had seven dining rooms installed in his Hollywood mansion. These rooms provide the structure for a 1970 collection of his recipes entitled *Liberace Cooks!*, authored by food critic Carol Truax. Readers follow Mr Showmanship from his regular dining room, where he shares recipes for pierogi and brains in black butter, to his buffet space (gazpacho, beef stroganoff), to his outdoor dining loggia, where you get a taste of his signature shish kebab (*see* page 139).

Maya Angelou – The late great poet made no apologies for publishing two cookbooks in her lifetime ("Writing and cookery are just two different means of communication," she told the *Guardian*). The second book *Great Food, All Day Long*, from 2010, focused on savoury meals that "were as good to eat at 8.30am as they were at 8.30pm. And yes," she added, "you can have fried rice for breakfast."

Gerard Depardieu – Unsurprisingly indulgent recipes from France's acting giant: *Gerard Depardieu: My Cookbook* (2005) focuses on domestic French cuisine (*boeuf bourguignon*, *moules marinières*) and features anecdotes about breakfasting on rabbit *en gelée* with country bread and glasses of cold white wine.

MORE FAMOUS COOKBOOK AUTHORS

Alicia Silverstone – The actor published *The Kind Diet: A Simple Guide to Feeling Great, Losing Weight and Saving the Planet* in 2009.

Sheryl Crow – The musician published *If It Makes You Healthy: More Than 100 Delicious Recipes Inspired by the Seasons* with her personal chef Chuck White in 2011.

Ted Nugent – The rocker and conservative activist published *Kill It & Grill It: A Guide to Preparing and Cooking Wild Game and Fish* in 2002.

Dolly Parton – The country singer published *Dolly's Dixie Fixin's: Love, Laughter and Lots of Good Food from My Tennessee Mountain Kitchen* in 2006.

Vincent Price – The actor published *A Treasury of Great Recipes* with his wife Mary in 1965.

Stanley Tucci – The *Big Night* star published *The Tucci Cookbook* in 2012 and *The Tucci Table* with his wife Felicity Blunt in 2015.

Terence Stamp – The actor and entrepreneur published *The Stamp Collection Cookbook* with Elizabeth Buxton in 1997.

2 Chainz – The rapper published *#Meal Time* in 2013.

Isaac Hayes – The soul legend published *Cooking with Heart & Soul* in 2000.

Yul Brynner – *The King and I* actor published *The Yul Brynner Cookbook: Food Fit for the King and You* in 1983.

A GLOSSARY OF FOOD PHOBIAS

Mine is ovophobia: I have an irrational fear of eggs, particularly the hard-boiled variety. Quite aside from the taste, I can't bear the sulphurous smell, or the pallid appearance, or the texture of the white, with its wobbly-smooth exterior and weirdly sandy inside. *Shudder!* Having always felt a bit silly about my egg aversion, I take some solace in the fact that other people have food phobias at least as odd as mine (*see* also page 118).

Fear of/aversion to…	Is known as…
Chicken	Alektorophobia
Garlic	Alliumphobia
Meat	Carnophobia
Fish	Ichthyophobia
Vegetables	Lachanophobia
Cooking	Mageirocophobia
Choking	Pnigophobia
Vomiting	Emetophobia
Poisoning	Toxiphobia
Dinner conversations	Deipnophobia
Peanut butter sticking to mouth	Arachibutyrophobia
Hot things	Thermophobia
Cold things	Frigophobia
Alcohol	Methyphobia
Fatty food	Lipophobia
Gaining weight	Obesophobia
Mushrooms	Mycophobia
Shellfish	Ostraconophobia
Chopsticks	Consecotaleophobia
Taste	Geumophobia
Sourness	Acerophobia
Drinking	Dipsophobia
Swallowing	Phagophobia
Food	Sitophobia

THE 15-YEAR HUNGER STRIKE

In November 2000, Indian activist and poet **Irom Chanu Sharmila** learned of the massacre of ten people in the northeastern state of Manipur, reportedly carried out by a government-controlled paramilitary group. In protest, she vowed to stop eating and drinking until a special powers act enabling state violence was repealed. She was 28 at the time. Three days later, Sharmila was arrested and charged with attempted suicide. In custody, she was force-fed through a nasal tube to keep her alive.

This is an extremely unpleasant procedure: when the rapper Yasiin Bey (formerly Mos Def) submitted to it in 2013, to draw attention to the abuse of prisoners in Guantánamo Bay, it reduced him to tears and he withdrew before it was finished. Sharmila was force-fed for nearly 16 years. Eventually the Iron Lady, as she was known, decided to call off her fast and engage more directly in local politics. On 9 August 2016, the world's longest hunger strike ended with a single lick of honey.

OTHER NOTABLE HUNGER STRIKERS

Bobby Sands (Northern Ireland, 1981) – Demanded that IRA members be treated as political prisoners. Was allowed to die after 66 days. Became a martyr to the Republican cause.

Guillermo Fariñas Hernández (Cuba, 2010) – Went 134 days without food after the death by hunger strike of a fellow dissident. Secured the release of 52 political prisoners.

Mia Farrow (USA, 2009) – Planned a three-week liquid-only fast to protest against the expulsion of aid agencies from Darfur by the Sudanese government. Stopped after 12 days citing health concerns. Richard Branson took over for an additional three days.

Cesar Chavez (USA, 1968) – Undertook several hunger strikes to draw attention to farm workers' rights, fasting for 25 days in 1968, 24 days in 1972, and 36 days in 1988 when he was 61.

ABSTEMIOUS EATERS

The motivations for eating simply and modestly, when one has access to luxuries of all kinds, are various and often complex. Usually health is the major consideration ("Eat food. Not too much. Mostly plants," is journalist Michael Pollan's boiled-down advice for healthy eating). But sometimes a disinterest in food, or even disgust at the idea of indulgence, is at the heart of the matter.

» **Steve Jobs** – "I was on one of my fruitarian diets [and] had just come back from the apple farm." This is Steve Jobs explaining the origins of the Apple company name to biographer Walter Isaacson and revealing something of his eating habits in the process. A vegan for most of his adult life, Jobs dabbled with the even more restrictive fruitarian diet and would spend weeks at a time eating only one or two foods, such as apples or carrots. "He believed that great harvests came from arid sources, pleasure from restraint," noted Jobs' daughter Lisa.

» **Henry David Thoreau** – The author of *Walden* professed little enthusiasm for food. "The wonder is how … you and I can live this slimy, beastly life, eating and drinking," he wrote. He avoided meat and alcohol. Coffee and tea were dangerous temptations. Salt he regarded as "that grossest of groceries". Cranberry-pickers were like butchers who "rake the tongues of bison out of the prairie grass". Even water was an indulgence he would gladly have shunned were he able to live without it.

» **David Bowie** – The Thin White Duke, Bowie's alter ego in the mid-1970s, was aptly named. According to biographer David Buckley, the singer was surviving at the time on a diet of red peppers, cocaine and milk slugged straight from the carton. (He did recover his appetite, however: one of his favourite dishes in later life was shepherd's pie.)

» **Jackie Onassis** – It's not austere, exactly, but legend has it that the former First Lady would eat one baked potato per day stuffed with Beluga caviar and soured cream. Disciplined about her weight, she watched the scales "with the rigor of a diamond merchant counting his carats," according to her social secretary Tish Baldrige. If she went a couple of pounds over, she would fast for a day, then confine herself to a diet of fruit until she was back to normal.

LIBERACE'S SHISH KEBABS

This recipe is taken from the fabulous *Liberace Cooks! Recipes from His Seven Dining Rooms* (1970) by Carol Truax. *See* page 134 for a more in-depth description of this and other collections of famous people's recipes.

Serves 8

¼ cup red wine

2 tbsp lemon or lime juice

3 tbsp olive oil

1 tsp salt

½ tsp freshly ground pepper

⅛ tsp oregano

2 cloves garlic, crushed

3lb lean lamb, cut into 1-inch cubes

1 eggplant or 2lb zucchini

2 green peppers, seeded and parboiled

4–6 firm tomatoes, cut into quarters

12–24 boiled or canned white onions

Mix the wine, lemon or lime juice, oil, salt, pepper, oregano, and garlic. Put the lamb into this marinade for several hours, turning frequently. Cut the unpeeled eggplant or zucchini into 2-inch pieces. Cut the peppers into eights. Alternate lamb and vegetables on 8 long skewers. Put extras on smaller skewers for seconds! Broil 8 minutes, turning once and brushing with the marinade.

QAT

In Yemen, around 40 per cent of the water supply goes towards irrigating a plant with no nutritional value. The appeal of *qat* or *khat* (*Catha edulis*) lies in its stimulant qualities. When chewed, or rather held in the cheek for long periods, the freshly cut leaves release an amphetamine-like chemical called cathinone, which causes euphoria and alertness. It now grows all over Africa, though the plant, and the practice of chewing it, has deepest roots in the Horn of Africa and the Arabian peninsula. It's estimated that 70–80 per cent of Yemenis between the ages of 16 and 50 have used it.

CLASSIC DESSERTS OF THE 21ST CENTURY

There's no clear formula for making a truly great dessert: it might involve several dozen processes of insane complexity, or else just a couple of clever tweaks and a catchy name. Here are four standout sweet dishes of the century so far which span both extremes.

Crack Pie
Inventor: Christina Tosi, Momofuku Milk Bar, New York, USA
Why it works: An oaty biscuit base with salty, sugary, buttery goo on top – what's not to love? As addictive as its name suggests.

Rubik's Cake
Inventor: Cédric Grolet, Le Meurice, Paris
Why it works: Aside from its supreme Instagrammability, this cube of 27 mini cubic cakes is packed full of inventive flavours – pairings include apricot/rosemary and cherry/tarragon.

Salted caramel honeycomb doughnut
Inventor: Justin Gellatly, Bread Ahead, London, UK
Why it works: Gellatly perfected the gourmet doughnut at his London bakery, a fact confirmed by this oozy salted caramel specimen pierced with a shard of honeycomb.

Botrytis Cinerea
Inventor: Heston Blumenthal, The Fat Duck, Bray, UK
Why it works: This wine-themed dessert is wildly complex, with some 80 components including yogurt, chocolate, peach, grape, vanilla and Roquefort powder, but they combine to stunning effect.

THE EDISON OF PUDDINGS

Born in the working-class city of Beauvais north of Paris, the French baker **Dominique Ansel** has become renowned for his mega-popular pastries and desserts, which circulate through the collective consciousness of sweet-lovers with the viral power of memes. His most hyped confection is the Cronut, a deathless marriage of croissant and doughnut – queues to buy Cronuts from his bakery in New York were known to form at 5am. But he has also brought us the Frozen S'more (vanilla ice cream covered in chocolate *feuilletine*, then enveloped in honey marshmallow and torched) and the Magic Soufflé, which claims to be the only soufflé that doesn't collapse, as well as reinventing the Breton *kouign-amann*. Ansel now has outlets in Tokyo and London (but not, at time of writing, Paris) and seems to be inexorably asserting his dominance over the great domain of desserts – he was named world's best pastry chef in 2017.

THE WORLD'S FAVOURITE CONFECTIONERY

Country	Product (company)	What is it?
USA	M&Ms (Mars)	Chocolate drops
Brazil	Trident (Cadbury's)	Chewing gum
UK	Dairy Milk (Cadbury's)	Milk chocolate bar
Germany	Milka (Mondelēz)	Milk chocolate bar
France	Hollywood (Mondelēz)	Chewing gum
Italy	Vivident (Perfetti Van Melle)	Chewing gum
Russia	Orbit (Wrigley)	Chewing gum
Czech Rep.	Orion (Nestlé)	Chocolate
China	Dove (Mars)	Chocolate
Japan	Meiji (Meiji Seika Kaisha)	Chocolate
Poland	Ptasie Mleczko (Cadbury's)	Chocolate/meringue
South Korea	Ghana (Lotte)	Milk chocolate
Denmark	Haribo (Haribo)	Gummy bears

THE GANNET EXPLAINS...
SMALL PLATES

What are they?
Taking a cue from tapas culture, many new restaurants have been rejecting the traditional starter/main/dessert formula in favour of smaller dishes which come out when they're ready and are designed to be shared with others. (The same restaurants are likely to have a no-reservations policy, with queues out the door, and feature small tables and/or counter space.)

What's the appeal?
For customers, you get to try more dishes without (a) incurring too much expense – small plates come with smaller prices – and (b) eating yourself senseless. Chefs like the set-up because it allows them to focus on each order as it comes in and send dishes out as soon as they're ready, without waiting for you to finish your starter. And restaurateurs are happy because they can turn tables quicker.

Ah, so it's just another way of milking the customers?
There's some validity to that claim. (It could also be levelled at the no-bookings policy, which means that another customer is ready to fill your seat the minute you've vacated it.) Those who despair at small plates also point out that they're often ill-suited for sharing, that they create a big pile-up of crockery on the (usually small) table, and that the rounded meal you get from a course-based menu is lost amid a cacophony of clashing flavours. They're not necessarily cheaper either: all those little price tags begin to add up after you've hungrily ordered an extra couple of dishes.

So we should be saying "no gracias" to small plates and demanding our main courses back?
Not so fast! There is real pleasure in being able to sample lots of different dishes – including things you've never tried before and may not have ordered otherwise. This arrangement also encourages chefs to be more adventurous: freed from the obligation to fill you up with a main, they can be more experimental. Whether or not this is a good thing very much depends on the calibre of the person cooking your dinner.

GREAT MEALS IN LITERATURE: DINNER

"I took the liberty of preparing a little snack for you," says Chojnicki, a Polish count, midway through Joseph Roth's 1932 novel **The Radetzky March**[1]. This is a fine understatement. We are in a gloomy hunting lodge on the desolate eastern border of the Austro-Hungarian Empire, which is heading towards collapse. The circumstances are bleak – and Chojnicki's dinner with our protagonist Carl Joseph Trotta and his father is overhung with an air of doom – but the spread itself is magnificent to behold.

The brown liver pâté, studded with pitch-black truffles, lay in a glittering wreath of fresh ice crystals. The tender breast of pheasant loomed lonesome on the snowy platter, surrounded by a gaudy retinue of green, red, white, and yellow vegetables, each in a bowl with a blue-gold rim and a coat of arms. In a spacious crystal vase, millions of pearls of black-gray caviar teemed within a circle of golden lemon slices. And the round pink wheels of ham, guarded by a large three-pronged silver form, lined up obediently in an oval bowl, surrounded by red-cheeked radishes that reminded one of small crisp country girls. Boiled, roasted, and marinated with sweet-and-sour onions, the fat broad pieces of carp and the narrow slippery pike lay on glass, silver, and porcelain. Round loaves of bread, brown and white, rested in simple, rustically pleated straw baskets, like babies in cradles, almost invisibly sliced, and with the slices so artfully rejoined that the bread looked hale and undivided. Among the dishes stood fat-bellied bottles and tall narrow crystal carafes with four or six sides and smooth round ones, some with long and others with short necks, with or without labels; and all followed by a regiment of glasses in various shapes and sizes. They began to eat.

See also…
» The *boeuf en daube* in Virginia Woolf's *To the Lighthouse* (1927).

[1] From Joachim Neugroschel's translation (1995).

BIZARRE ICE CREAMS

"I scream, you scream, we all scream for... cow's tongue ice cream?" Perhaps not, but that doesn't stop certain adventurous parlours experimenting with crazy flavours – and even more adventurous ice-cream lovers around the world from buying them.

What?	Where?	Why?
Crocodile egg	Sweet Spot, Davao City, Philippines	More nutritious than chicken eggs, say owners
Breast milk	The Icecreamists, London (now closed)	"What could be more natural than fresh, free-range mother's milk in an ice cream?" said a donor
Coronation chicken	Gelupo, London	To celebrate the Queen's diamond jubilee
Haggis	Glen Urr, Dumfries, Scotland	For patriotic reasons, I assume
Salmon & cream cheese	Max & Mina's, Flushing, NYC	A customer suggested it and they gamely followed through
Cheeseburger	Heladeria Coromoto, Merida, Venezeula	They have over 800 flavours, so why not one more?

Japan is pretty much the world leader in weird ice-cream flavours. Here are some of the delights on offer:

Deep-fried oyster	Chicken wing	Curry	Crab
Shark fin	Cactus	Whitebait	Octopus
Raw horse	Python	Squid ink	Fish
Eel	Jellyfish	Beer	Sweet potato
	Cow's tongue	Stew	Shrimp
			Charcoal

FOOD HOTSPOTS:
THE HISTORIC DINING ROOM

La Tour d'Argent, Paris – If you believe the in-house history, this grand restaurant overlooking Notre Dame began life as an inn way back in 1582 and was frequented by Henry IV of England, who "came regularly to savour the heron *pâté*". Service was interrupted by the French Revolution in 1789 but snapped back into action a decade later under Napoleon's private chef, M. LeCoq. Since the 1890s, La Tour has been famous for its pressed duck, and the serial numbers accorded to each bird have also catalogued some of the restaurant's famous visitors: #112,151 went to Franklin D Roosevelt, #203,728 to Marlene Dietrich, and #253,652 to Charlie Chaplin.

Stiftskeller St Peter, Salzburg – A contender for oldest restaurant in the world, if its mention in a text by Alcuin of York in the year 803 can be taken at face value, this venerable inn within the walls of St Peter's Abbey undoubtedly has history. Conflicts have gusted through – French troops were given quarters here during the Napoleonic wars – as have many famous names. Where else can count Mozart and Clint Eastwood, Christopher Columbus and Bill Clinton among its clientele?

Aragvi, Moscow – Opened in 1938, Moscow's first Georgian restaurant was a glitzy hangout for Soviet bigwigs, playing host to cosmonauts, chess champions and KGB agents. Stalin's murderous chief of secret police, Lavrentiy Beria, was a regular customer, as was British double-agent Kim Philby. Its fortunes declined in tandem with the fall of the Soviet Union, but a recent $20m renovation has brought Aragvi back to opulent life.

OTHER RESTAURANTS WITH STORIED PASTS

Rules, London – Hosted Charles Dickens and Edward VII, has featured in novels by Graham Greene and Evelyn Waugh.

Ma Yu Ching's Bucket Chicken House, Kaifeng – China's oldest restaurant, weathering wars and dynasties since 1153.

SILICON VALLEY IS
DISRUPTING YOUR MEAL

Having shaken up the hotel, taxi, music, publishing and retail industries, to name but a few, Silicon Valley is now turning its attention to food. A stampede of tech startups is threatening to revolutionize major areas of food production, from meat to dairy to the entire meal – and investors have been pumping hundreds of millions of dollars into their endeavours.

Meat

Who? Impossible Foods (also: Beyond Meat, Memphis Meats)

What? Their high-profile product is a burger which "looks, cooks, smells, sizzles and tastes like conventional ground beef but is made entirely from plants". The "magic ingredient" that gives it a meaty flavour is a soy-based protein called leghemoglobin, aka "heme".

Why? To reduce the environmental and health impacts of eating meat while retaining (at least some of) the satisfaction.

Plus: According to the company, the Impossible Burger uses 95 per cent less land and 74 per cent less water and emits about 87 per cent less greenhouse gases than a burger from cows. Top New York chef David Chang added it to his menu at Momofuku Nishi in 2016.

Minus: Production at the moment is limited (though the company says they'll soon be producing up to 454,000kg (1,000,000lb) of meatless meat per month).

Dairy

Who? Perfect Day (formerly Muufri)

What? "Animal-free milk" made by fermenting dairy yeast and sugar, then adding plant fats, nutrients and proteins that can be found in milk.

Why? To allow vegans, environmentalists and the dairy intolerant to enjoy products that more closely resemble dairy.

Plus: Their "milk" is high in protein and lactose free, and they estimate it causes 84 per cent less greenhouse emissions and 98 per cent less water consumption than real milk.

Minus: Dabbling in "cellular agriculture" – producing meat and animal products through cell cultures rather than livestock – may turn some consumers off.

Eggs

<u>Who?</u> Hampton Creek

<u>What?</u> Egg-based products such as mayo, but without the eggs.

<u>Why?</u> To alleviate the plight of battery hens and reduce the carbon footprint of the egg industry.

<u>Plus</u>: The company claims it has been valued at $1.1bn since launching in 2011, showing there's an appetite for egg-free alternatives.

<u>Minus</u>: Hampton Creek was accused of exaggerating the benefits of its products and misrepresenting their shelf life. It also allegedly bought its own product off the shelves to create the appearance of higher demand.

Coffee

<u>Who?</u> Bulletproof Coffee

<u>What?</u> "Low-toxin" coffee blended with butter and Brain Octane Oil, a triglyceride oil derived from coconuts.

<u>Why?</u> Founder Dave Asprey was inspired by Tibetan yak butter tea, which he said made him "feel amazing".

<u>Plus</u>: Asprey's concoction is designed to boost concentration and improve general health.

<u>Minus</u>: There have been mixed reports about the taste and oily mouthfeel, and some doubts have been cast on the health claims that accompany it.

The entire meal

<u>Who?</u> Soylent (also: Ambronite, Space Nutrients Station, Huel)

<u>What?</u> Soylent's main product is a powder containing a range of nutrients (including maltodextrin, soy protein and sunflower oil) which is mixed with water to create a meal replacement drink.

<u>Why?</u> To cut the time and money spent buying and preparing food.

<u>Plus</u>: Contains no animal products. They claim one 400ml (14fl oz) bottle of their pre-made drink yields 20 per cent of your daily nutritional requirements.

<u>Minus</u>: Flavour is not Soylent's selling point (one early critic likened the powdered drink to watered-down semen). Two products were withdrawn in October 2016 after some customers reported gastrointestinal problems.

THE CHEF WHO COOKED
FOR KIM JONG-IL

In 1982, a Japanese chef known by the alias **Kenji Fujimoto** moved to North Korea to teach sushi skills at a Pyongyang cookery school. Six years later, he became personal chef to supreme leader Kim Jong-il, a post he held until his escape from the country in 2001.

The books he's written about those years abound with tales of Kim's lavish appetites, which contrasted with the extreme poverty of the country he ruled over. Here are a few examples…

» The first time Fujimoto worked for Kim Jong-il, he served him *toro* (bluefin tuna), which went down well. Thereafter Kim would call out to the chef on a regular basis: "*Toro*, one more."

» Fujimoto impressed Kim's coterie with a technique he picked up while working at Tokyo's Tsukiji fish market (*see* page 178): he took a live fish, filleted it around the organs, and served it – still living – to the leader and his entourage.

» At his official residence, Kim had a liquor cellar containing some 10,000 bottles – his favourite brands included Johnnie Walker Swing and Hennessy XO. The cellar also housed a karaoke set and a piano, and it was here that Fujimoto met Om Jong-yo, a North Korean pop star whom Kim summoned to sing for the chef's pleasure. Kim then arranged for Fujimoto to marry Om, which he dutifully did in the late 1980s.

» As part of his job, Fujimoto was often dispatched around the world to pick up delicacies unavailable in North Korea. Here's an overview of his shopping list:

Japan...........................Seafood
France..........................Wine, Cognac
Denmark.......................Pork
Iran/Uzbekistan..............Caviar
Czechoslovakia...............Draft beer
Urumqi (NW China)..........Fruit (hamigua melons, grapes)
Thailand/Malaysia...........Fruit (durians, papayas, mangoes)

» Kim once sent Fujimoto to Japan to buy rice cakes filled with mugwort, as well as every brand of Japanese cigarette. Each cake cost around 100 yen, but factoring in travel and accommodation Fujimoto calculated that the unit cost rose fifteen-fold. Kim smoked only the menthol cigarettes and proclaimed the rice cakes "really delicious".

» Kim set up an institute dedicated to his longevity, with a staff of 200 monitoring every detail of his diet. Only perfectly shaped North Korean rice made it onto the leader's plate, each grain hand-inspected for defects. The rice, according to Fujimoto, had to be cooked over wood gathered from the sacred Mount Paektu.

» Kim was particularly fond of Japanese cooking shows such as *Iron Chef* and *Which Dish?*. After watching an episode of the latter featuring *uni* (sea urchin roe), Kim sent Fujimoto to fetch some from Japan, despite the fact that the chef had been caught spying for the Japanese and was banned from travelling. Fujimoto gave his minders the slip at Tsukiji fish market.

» In 2012, a year after Kim Jong-il's death, Fujimoto returned to Pyongyang on the invite of Kim Jong-un. Rather than having the man who betrayed his father killed, the young leader challenged Fujimoto to a drinking contest. The chef has since returned to live in Pyongyang, where he now runs a sushi restaurant called Takahashi.

RAGÙ

There are many regional versions of this meaty pasta sauce (not to be confused with *ragoût*, a French stew of meat and vegetables, though the names are connected). Outside Italy, however, it is virtually synonymous with Bolognese. Aghast at how the dish was being corrupted – with such excrescences as turkey and cream – the *Accademia Italiana della Cucina* responded in 1982 by setting down an "authentic" version – ground beef, pancetta, onion, carrots, celery, tomato paste, milk, white or red wine, with tagliatelle instead of spaghetti – in a bid to restore the dish's integrity.

MY COMFORT FOOD

One thing I always want to find out when I do interviews for *The Gannet* is how people eat on their days off. What do you cook when you're not trying to impress anybody and simply want a comforting meal? It's a question that comes up again and again, and the answers are often quite revealing.

"Sometimes I really want to eat this garlic soup that my mum and grandmother make. It's basically just garlic, vermicelli, eggs and vinegar and it's really nice, but it's more comforting when I'm not the one making it, do you know what I mean? It's better when my mum makes it." – **Alice Quillet, chef, Paris, France**

"Chicken-fried steak, white gravy, macaroni and cheese, mashed potatoes and carrots. Did I say mashed potato? [laughs]" – **Jack Gilmore, chef, Austin, TX, USA**

"If it's something other than toast (I eat an awful lot of toast with various toppings) then it would have to be a grain risotto. It could be barley, orzo, spelt or freekeh cooked with veggies and spices, lots of ginger and tamarind, and finished with heaps of grated Manchego, Parmesan or pecorino – whatever is in the fridge." – **Peter Gordon, chef, London, UK**

"Beef rouladen [a stew with bacon, mustard and gherkins] is something I love cooking in the colder months. It's a German dish that my father used to make. He used to sit at the table with a bottle of wine while preparing it and we'd help him – with the wine and with battering out the beef – amid a lot of chatter. It's a simple and really delicious dish [the recipe is on page 182]." – **Laragh Stuart, food producer, Wicklow, Ireland**

"Ice cream. The one I go for, if I'm just buying from the corner shop, is Häagen-Dazs strawberry cheesecake. And I love Poco Gelato [in Leigh-on-Sea, Essex] – his flavours are incredible. I should have been an ice-cream maker, but that would kill the enjoyment for me, wouldn't it?" – **Kylee Newton, founder of Newton & Pott preserves, London, UK**

"A bowl of ramen for me is the ultimate comfort food. Having lived four years in the snowy north of Japan, I really appreciate all different types of ramen [*see* page 122]: the rich bone broths, the light salt or soy broths, the thick miso broths of Hokkaido. Decent ramen is not readily available in Copenhagen so it's one of the first things I have to have when I arrive in Tokyo. Afuri is a favourite restaurant: I just love the addition of yuzu to the bowl and will always add extra toppings of soft egg, roast pork and nori." – **Katherine Bont, Noma team leader, Copenhagen, Denmark**

"I really want to say that [my wife and business partner] Manuela makes the best Bolognese. We always have some in the freezer. Coming home: Bolognese, TV, black pumpkin-seed salad. It reminds me always of my childhood, what my mum would make." – **Konstantin Filippou, chef, Vienna, Austria**

"When we used to travel from Cape Town to Norway to visit my grandparents, it would take 24 hours door-to-door and I'd be so exhausted, but I knew that when we arrived, my favourite meatballs with potatoes, cucumber salad and my grandmother's gravy would be waiting for me. It was the same every year. That's my comfort food for sure." – **Julia Wakeham, designer, Cape Town, South Africa**

"If I'm alone and just want something simple, Nigel Slater's the one for me. He's got this pasta recipe where he confits 15 cloves of garlic in olive oil till soft, then has it with spaghetti, goats' cheese and thyme – just stir it all through. It's easy and I love the garlic and goat thing." – **Olia Hercules, food writer, London, UK**

"One day my comfort food could be jerk chicken, rice and peas. The next it could be a Persian stew or a shepherd's pie. It's usually something involving carbs. We're not afraid of carbs in our house. Or butter." – **Sabrina Ghayour, food writer, London, UK**

"Frozen peas. I just love them. I always have bags of them in the freezer. When I'm working really hard, I'll have a bowl of peas with olive oil and black pepper when I come home." – **Katie Sanderson, chef, Dublin, Ireland**

FOOD IS THE NEW ROCK

Will cooking, at its most aestheticized and revolutionary, ever be as cool as music-making? Perhaps not, though the gap has certainly narrowed in recent years. It's now a cliché to compare the latest hot chef to a rock star, though their tattoos and hairstyles are often the same and the levels of hysteria they excite among their fans reach a similar pitch. It's telling that musicians have been paying more attention to food of late, and that some have even been getting in on the game...

Action Bronson's food show
Presenting a programme called *Fuck, That's Delicious* for Viceland was not a total shot in the dark for Action Bronson: the larger-than-life rapper started out as a chef in his native New York. He certainly isn't your average food show host, punctuating his dining experiences around the world with profanity, casual drug use and vigorous stage performances.
See also:
» Franz Ferdinand lead singer Alex Kapranos's on-tour food journal for the *Guardian* in 2005–6
» The Roots drummer Questlove's 2016 book *Something to Food About: Exploring Creativity with Innovative Chefs*

Kenny Chesney's rum
It's unlikely that the Tennessee country music singer actually rolls up his sleeves and makes Blue Chair Bay rum, which he launched in 2001 and is now available in seven varieties (including coconut spiced and banana rum cream), but the company does insist "he makes a final call on everything" right down to the wood used in the cork.
See also:
» Drake's Virginia Black whiskey
» Marilyn Manson's Mansinthe absinthe
» Justin Timberlake's Sausa 901 tequila

Smokey Robinson's microwaveable bowls
"The gumbo *is* a gift from God," declared the Motown legend in 2004, announcing a new addition to his line of microwaveable frozen meals.

Besides "The Soul is in the Bowl" gumbo, the range included jambalaya and red beans and rice.
See also:
»   Sylvester Stallone's High Protein Pudding
»   Kiss Frozen Rockuccino coffee, Kiss Destroyer beer and Kiss Hotter Than Hell ketchup – all endorsed by the New York shock-rockers
»   George Foreman's Lean Mean Fat-Reducing Grilling Machine

Kelis's restaurant pop-up
In July 2016, New York R&B artist and Cordon Bleu graduate Kelis stepped up her love of food – already evident on albums such as, yes, *Food* – and did a four-day kitchen residency at London's Leicester House. There, she dished up such delights as truffle aji cheeseburgers and arepas with spicy tuna, sour mango and scotch bonnet.
See also:
»   London rapper Loyle Carner's cooking class for teenagers with ADHD, pleasingly entitled Chilli Con Carner

Ziggy Marley's cookbook
The title – *Ziggy Marley and Family Cookbook: Delicious Meals made with Whole, Organic Ingredients from the Marley Kitchen* – is fairly self-explanatory. The reggae star, son of Bob, launched a line of hemp seeds and flavoured coconut oils in 2013 and was duly inspired to write a cookbook "based on the foods of his life". From this we may infer that his life involves lots of healthy juices and salads alongside Jamaican classics such as jerk chicken and escabeche fish. For more examples of celebrity cookbooks, *see* pages 134–5.

STAGIAIRE

Many of the world's top restaurants share a secret ingredient: the stagiaire, or unpaid kitchen intern. Noma in Copenhagen usually had up to 30 interns working alongside 25 full-time kitchen staff.
Despite the lack of pay, many young chefs deem it worthwhile for the prestige and the skills they pick up. It's worth it for their employers too: without stagiaires, many top restaurants would struggle to make it to the end of service.

GOURMANDS OF PAGE AND SCREEN

There are plenty of books and films in which food plays a starring role, but fewer where the main character is a bona fide foodie – and fewer still where the protagonist displays their food knowledge without tipping over into egregious snobbery. Let's assess some of fiction's most fêted fine-diners…

James Bond
Some 500 meals are mentioned in Ian Fleming's oeuvre. "I take a ridiculous pleasure in what I eat and drink," admits the MI5 spy in *Casino Royale*.
Likes: For breakfast – "Bond's favourite meal of the day," as described in *From Russia with Love* – a single boiled egg from French Marans hens with two slices of wholewheat toast, a pat of Jersey butter and conserves including Norwegian heather honey; and two cups of very strong coffee from De Bry on New Oxford Street, brewed in an American Chemex.
Dislikes: Tea. "I hate it," he rants in *Goldfinger*. "It's mud. Moreover, it's one of the main reasons for the downfall of the British Empire."
Gourmand ranking: 8/10
Snobbery level: High

Inspector Montalbano
The Sicilian detective, created by Andrea Camilleri, describes food as "the accelerator of my brain's functioning system" in *La Caccia Al Tesoro* and never skips a meal during a case unless something's badly wrong.
Likes: Seafood (a sauté of clams in breadcrumbs, spaghetti with white clam sauce, and a roasted turbot with oregano and caramelized lemon, leaves him "deeply moved" in *The Snack Thief*). White wine. Anything made by his housekeeper Adelina, such as *pasta 'ncasciata* followed by rabbit *alla cacciatore* (which brings tears of happiness to his eyes in *The Potter's Field*) or a plate of cold pasta with tomatoes, basil and black *passuluna* olives dug out of his fridge (in *The Terracotta Dog*) – or Adelina's celebrated *arancini*.
Dislikes: Idle talking while eating. Picnics. He tends to avoid sweet things and usually forgoes breakfast in favour of several cups of espresso on his veranda.
Gourmand ranking: 9/10
Snobbery level: Low

Hannibal Lecter

Not just a culture vulture with a weakness for human flesh, Thomas Harris's creation is also a showy gourmand, whom we first encounter in a hospital for the criminally insane with a copy of Alexandre Dumas's *Grand dictionnaire de cuisine* open on his chest.

Likes: *Pâté de foie gras*. Anatolian figs. A census taker's liver with fava beans and a big Amarone (downgraded to a "nice Chianti" in the film version of *Silence of the Lambs*).

Dislikes: Bad manners. "Whenever feasible, one should eat the rude," he muses in the *Hannibal* TV series.

Gourmand ranking: 4/10 (points deducted for cannibalism)

Snobbery level: Through the roof

Babette Hersant

Fleeing from the French revolution, the hero of the film *Babette's Feast* finds refuge in a remote Lutheran community in Denmark (the original Karen Blixen story places it in Norway). After 14 years of cooking simple meals for two sisters, she wins the lottery and blows her winnings on an elaborate dinner for the community.

Likes: Her seven-course feast in the film gives a good indication: *Potage à la Tortue* (turtle soup) served with Amontillado sherry; *Blinis Demidoff* (buckwheat pancakes with caviar and soured cream) served with Veuve Cliquot Champagne; *Cailles en Sarcophage* ("entombed" quail in puff pastry shell with foie gras and truffle sauce) served with Clos de Vougeot Pinot Noir; an endive salad; *Savarin au Rhum avec des Figues et Fruit Glacée* (rum sponge cake with figs and candied fruit) served with Champagne; assorted cheeses and fruits served with Sauternes; coffee with Vieux Marc Grande Champagne Cognac.

Dislikes: Bland Lutheran food, which she attempts to enliven over the course of the story.

Gourmand ranking: 9/10

Snobbery level: Medium

MORE FICTIONAL FOOD OBSESSIVES

Rachel Samstat in *Heartburn* by Nora Ephron

Tarquin Winot in *The Debt to Pleasure* by John Lanchester

Sirine in *Crescent* by Diana Abu-Jaber

THEIR FAVOURITE FOOD

Unlike his predecessor Lenin, who had very little interest in food, Joseph Stalin was a gourmand who issued many culinary demands during his 30 years in charge of the Soviet Union. As well as loving *nelma* (a Siberian whitefish) and wines from his native Georgia, he had a particular fondness for bananas and, according to one biographer, "got very cantankerous" when served a bad one. Here are some other notable figures and their culinary weaknesses.

VIP	Liked to eat
Martin Luther King	Sweet potato pie
Catherine the Great	Boiled beef with pickled cucumbers
Emperor Tiberius	Melons, cucumbers
Ferdinand Marcos	Malunggay (aka horseradish tree)
George IV	Peaches (especially stewed in brandy)
Kim Jong-un	Sushi, steak, Emmenthal cheese
Pope Clement VII	Mustard
Percy Shelley	Bread
Buster Keaton	Lobster Joseph
Pablo Neruda	Dulce de alcayota (spaghetti squash jam)
Sonia Rykiel	Chocolate
Charlton Heston	Peanut butter
Garrison Keillor	Sweetcorn[1]
Bruce Lee	Beef with oyster sauce
Gioachino Rossini	Foie gras, truffles
Clark Gable	Raw onions
Bette Davis	Potatoes
Kurt Cobain	Kraft macaroni and cheese
Cole Porter	Fudge

[1] According to the American author and radio broadcaster, "People have tried and they have tried, but sex is not better than sweetcorn."

WHAT WAS EATEN
AT THE LAST SUPPER?

The truth is, we don't really know. According to the Gospels, there was bread and wine, which Jesus claimed were his body and blood, and possibly meat, though scriptural references to Passover lamb are far from clear. In the absence of firm historical evidence, let's turn to the realms of art to see how different ages have viewed Christianity's most famous repast.

Fish – Often taking pride of place in Late Antique and Byzantine depictions, fish is the main attraction in the earliest surviving representation of the Last Supper: a 6th-century mosaic at Sant'Apollinare Nuovo in Ravenna, Italy.

Fruit – Most paintings of the Last Supper (even Salvador Dalí's 1955 version) portray a rather austere meal. Not so Alessandro Allori's 1582 fresco in Chiesa di Santa Maria del Carmine, Florence, which depicts a veritable feast of berries, apples and pears alongside the usual bread and wine.

Lamb – Despite scriptural ambiguity, numerous artists – including Albrecht Dürer and Lucas Cranach the Elder – have placed a whole lamb at the centre of the table. In Jacopo Bassano's 1546–8 painting, Christ fingers a gruesome platter bearing a lamb's disembodied head.

Grilled eels – After Leonardo da Vinci's famous Last Supper mural in Milan was cleaned up in 1997, culinary details emerged. According to an article in the journal *Gastronomica*, what we're seeing on the far right of the table are plates of sliced grilled eel with orange. This, the author believes, has less to do with biblical accuracy than the popularity of eels in Renaissance Italy and a contemporary fad for pairing fish with orange.

Roast chicken – British artist Yinka Shonibare doesn't just mess with culinary conventions in his 2013 sculptural tableau of the Last Supper, shoehorning in pineapples, mussels and bottles of Moët alongside the roast chicken. He also tweaks the cast of characters, replacing Christ with a goatish Bacchus and the apostles with headless mannequins.

THE MOST EXTRAVAGANT FEAST IN FICTION?

In Petronius's *Satyricon*, written in Rome during the reign of Nero, we are plunged into a wildly ostentatious dinner party at the home of Trimalchio, a wealthy former slave who embodies to bursting point the concept of *nouveau riche*. It's everything you'd imagine Roman gastronomic excess to be and more, though in its wacky playfulness the food isn't a million miles away from what you'll find on pages 7 or 48–9.

I
A bronze donkey bearing panniers of olives flanked by
hot sausages, damson plums
and dormice with poppy seeds and honey

II
A wooden hen on a nest of peahens' eggs,
which contain figpeckers embedded in peppered egg yolk

III
Twelve dishes arranged upon a zodiacal tray
which lifts to reveal stuffed capons, sows' udders
and a hare with wings

IV
Whole wild boar served with baskets of Syrian and Theban dates
and pastry piglets filled with live thrushes

V
A pig stuffed with sausages and blood puddings

VI
A boiled calf wearing a helmet, slashed to pieces by a slave dressed as
Ajax who serves the meat with his sword

VII

A pastry likeness of the fertility god Priapus bearing
fruits and cakes which burst into clouds of saffron

VIII

Fat capons and goose eggs with pastry hoods

IX

Pastry thrushes stuffed with raisins and nuts, accompanied by quinces
spiked with thorns to resemble sea urchins

X

A fat goose surrounded by fish and game,
all of which turn out to be made of pork

XI

Water jugs which are broken to reveal oysters and scallops,
followed by a gridiron of snails

XII

A freshly killed cock (which crowed early – a bad omen according to
Trimalchio) pot-roasted in wine and served to the guests before they
make good their escape

TRIMALCHIO IN WEST EGG…

…was a working title for F Scott Fitzgerald's 1925 novel *The Great
Gatsby*. The reference is not accidental. Both Gatsby and Trimalchio
have risen from the bottom rungs of society to occupy positions of
fabulous wealth. Both are viewed with suspicion, even disgust, by the
upper classes who feast at their expense. And both have a fine line in
show-stopping dinner parties. At Gatsby's mansion in West Egg
(a fictional town on Long Island), Champagne flows, cocktails float
about and carousing guests are met by spectacles such as this:

> On buffet tables, garnished with glistening hors d'oeuvre,
> spiced baked hams crowded against salads of harlequin
> designs and pastry pigs and turkeys bewitched to a dark gold.

A DISH BEST SERVED COLD

Revenge as a theme is not exactly underrepresented in world literature. Throughout history, authors have had fun devising inventively nasty ways for one character to get their own back on another. Some of the most shocking – because they're so integral to our everyday lives – involve food and drink. How better to catch out your enemy than when they're thinking with their stomach?

Titus Andronicus (William Shakespeare, 1593)
The offence: Titus, a general in the Roman army, is locked in a cycle of revenge with Tamora, queen of the Goths, which reaches its awful nadir when Tamora's two sons and lover Aaron rape Titus's daughter Lavinia.
The payback: Titus kills the two sons and bakes them into a pie, which he serves to their unsuspecting mother at a feast. (This scene is echoed in season 6 of the HBO series *Game of Thrones*, when an unsuspecting father tucks into a similar type of pie.)
Inventiveness: 9/10

The Cask of Amontillado (Edgar Allen Poe, 1846)
The offence: An unspecified insult that an Italian nobleman named Fortunato has visited upon the narrator Montresor.
The payback: Montresor lures Fortunato down to his family vaults during the carnival to taste a newly acquired pipe (491 litres/130 gallons) of rare Amontillado sherry, plying him with wine along the way. Then, telling his inebriated "friend" that the Amontillado lies within a remote recess, he claps Fortunato in chains and seals him into the recess forever with bricks and mortar.
Inventiveness: 6/10

American Psycho (Bret Easton Ellis, 1991)
The offence: The annoyingness of Evelyn Williams as perceived by her boyfriend, the novel's psychopathic narrator Patrick Bateman.
The payback: Bateman steals a urinal cake, covers it in chocolate syrup, places it in a Godiva box and arranges for it to be served to Evelyn for dessert at a "super-chic nouvelle Chinese restaurant". She very reluctantly eats the whole thing.
Inventiveness: 2/10 (points deducted for sheer unpleasantness)

Charlie and the Chocolate Factory (Roald Dahl, 1964)
The offence: Four of the five children who win access to Willy Wonka's factory get ejected according to their flaws: Augustus is gluttonous, Violet is obsessed with chewing gum, Veruca is greedy and manipulative, and Mike is addicted to TV.
The payback: Augustus falls into a river of chocolate; Violet blows up into a giant blueberry; Veruca is thrown down a garbage chute by nut-testing squirrels; and Mike gets shrunk by television.
Inventiveness: 8/10

Snow White (The Brothers Grimm, 1812)
The offence: When the titular princess grows up, she is deemed "the fairest in the land" by a magic mirror belonging to her stepmother the queen, who doesn't enjoy being relegated to second place.
The payback: After making several attempts on the princess's life, the queen tricks her into eating a poisoned apple. Snow White is presumed dead until roused by a passing prince.
Inventiveness: 3/10

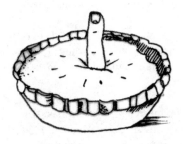

T

TEFF

The most important food crop in Ethiopia, where it has been farmed since prehistoric times, teff (or tef) is a grain the size of a poppy seed which, ground into flour, is used to make the spongy *injera* flatbread. Recently, teff has caught on in the West, where it's been heralded as a superfood due to high levels of calcium, iron and protein, and the fact that it is naturally gluten free.

WHAT THE WORLD'S STRONGEST MAN EATS IN A DAY

Brian Shaw of Fort Lupton, Colorado, has won the World's Strongest Man competition four times since 2011. To stay competitive and fuel his fearsomely intensive training sessions, he has to eat almost constantly from morning to night, "getting in" huge meals at hourly intervals. (Almost half his recommended daily calorie intake comes from breakfast alone.) What he eats is almost entirely dictated by a dietician. "I don't ask questions," says the genial strongman. "I just eat – and most of the time I get a smile at some of the stuff that I gotta eat."

Here's what Shaw consumes on an average day:

Meal #1: A bowl of Cinnamon Toast Crunch cereal. Eight eggs (scrambled). One heaped dessertspoon of peanut butter. *Calories: 1,180*

#2: Whey protein shake (80g/2 ¾ oz). Granola bars. Peanut butter. *Calories: 1,053*

#3: A huge bowl of minced organic grass-fed beef (fried) with angel-hair pasta and red sauce. *Calories: 2,190*

#4: Whey protein shake (80g/2 ¾ oz). Peanut butter. Frozen organic blueberries. *Calories: 1,002*

#5: Minced organic turkey with white jasmine rice and broccoli. *Calories: 1,417*

#6: A big takeaway meal with various types of pasta and sauce. A pint of cola with ice. *Calories: around 3,400*

#7: 4 slices of cheesecake, whey protein shake (80g/2 ¾ oz). *Calories: 1,649*

Total intake:
Calories: 12,019
Protein: 705g (1lb 9oz)
Carbohydrates: 1402g (3lb 1oz)
Fat: 399g (14oz)

THE GASTRONOMIC STOCK MARKET

What makes a certain type of food exorbitantly expensive at one moment in history and cheap as chips the next? Usually it comes down to basic supply and demand: we crave what we cannot easily have and disregard it once it becomes readily available – or vice versa.

RISING

Oysters – "Poverty and oysters always seem to go together," remarks Sam in Charles Dickens's *The Pickwick Papers* (1837). It's hard to imagine saying that now. Falling stocks hit by pollution and over-consumption are largely responsible for the oyster's rise in value.

Lobster – In some early American penal colonies, it was considered inhumane to feed lobsters to inmates more than once a week, so low was the arthropod's reputation in pre-modern times.

Foie gras – A delicacy in Ancient Egypt, fattened goose or duck liver has see-sawed in popularity over the centuries. It was considered a peasant dish in medieval Europe before finding favour with the French during the Renaissance.

FALLING

Black pepper – So great was the value of *Piper nigrum* in the first millennium that Alaric the Goth demanded 3,000 pounds (1.36 tons) of it as part of a ransom for Rome in 408. A key commodity in the spice trade, its value fell in the later Middle Ages due to oversupply.

Tea – Prior to becoming Britain's national drink around the mid-18th century, tea was such a rarity that only royalty and aristocracy could regularly afford it. Now tea bags can be bought for less than a penny.

Sugar – "A most precious product, very necessary for the use and health of mankind," is how one Crusade chronicler described sugar in the 12th century. Times have changed: sugar became cheap due to widespread cultivation and its health benefits are now very much in question.

THE GANNET EXPLAINS...
CLEAN EATING

What is it?
Clean eating is the practice of eating whole or "real" foods, as opposed to foods that have been heavily processed or refined, or that contain high levels of sugar, salt and fat, in order to promote general health and "wellness".

It'll never catch on.
Actually the clean-eating movement has enjoyed incredible popularity in recent years. Fronted mainly by attractive young people who claim to have boosted their health or even cured themselves of illness by changing their diet, it has taken the media by storm. London's "queen of green" Ella Mills (aka Deliciously Ella) boasts 1.1 million followers on Instagram and sold over a quarter of a million copies of her debut cookbook – although she now distances herself from the clean eating label.

Why, what's wrong with eating healthily?
Nothing as such. The trouble arises when it becomes a doctrine preached to an impressionable audience already prone to feeling bad about their bodies. To make matters worse, most of the people doing the preaching lack proper medical qualifications and often end up dishing out dodgy, scaremongering advice.

Can you give an example?
Many in the clean-eating movement have advised their readers to avoid gluten, when in fact only a very small percentage of the population is adversely affected by it. More disturbing is the case of Australian blogger Belle Gibson, who claimed to have cured herself of various cancers largely through eating the right foods. She made a lot of money by promoting her methods over conventional medical therapies, before it was revealed in 2015 that her entire story had been made up.

That's a pretty extreme case.
Yes, but it shows how dangerous it can be when unqualified, unregulated individuals with a powerful media voice start dominating the debate on how we should eat.

WHEN DOES LAMB BECOME MUTTON?

The question is less straightforward than it may appear. In Commonwealth countries, a lamb is a domestic sheep under a year old that does not have any permanent incisor teeth "in wear" – and lamb is the meat it produces. After this, the sheep and its meat are known as hogget, and when it has more than two permanent incisors in wear, it becomes mutton. In the USA, lamb skips straight to mutton after the 12-month mark, though mutton appears very rarely on American tables. In South Asia, to confuse matters further, "mutton" commonly applies to goat meat.

Let us extract ourselves from this tangle of definitions and look at how age is used to categorize other meat-producing animals[1].

Chickens – Broiler chickens raised for meat production are usually slaughtered between five and fourteen weeks. Hens start to lay eggs at around twenty weeks. The natural life expectancy of a chicken is between six and twelve years (smaller breeds generally live longer).

Pigs – Pigs are slaughtered at different ages: two to six weeks for suckling pigs; six weeks to three months for feeder pigs; four to six months for porkers, intended for pork; eight to ten months for baconers, intended for bacon. A boar is any male pig over six months old that's intended for use in the breeding herd.

Cows – Veal comes from calves of either sex aged up to eight months old, after which the meat is known as beef. There are several sub-categories of veal:
– bob veal: up to one month old
– milk-fed veal: around 18–20 weeks
– red or grain-fed veal: 22–26 weeks
– rose veal: 35 weeks

[1] Weight, however, is often more significant than age in deciding when animals get the chop.

DISTINCTIVE DISHES OF AFRICA

i. Pastilla (Morocco) – Aka *b'stilla*, this delicious pie traditionally contains pigeon meat (though chicken is commonly used), almonds, sugar and spices – all wrapped in paper-thin pastry.

ii. Ful medames (Egypt) – As old as the pharaohs by some reckonings, this dish of slow-cooked fava beans is typically enjoyed for breakfast with chopped parsley, chilli, onions and boiled eggs.

iii. Kitfo (Ethiopia/Eritrea) – Think of it as Ethiopian steak tartare: minced raw beef marinated in *mitmita* spice blend and *niter kibbeh* (a spiced clarified butter), usually eaten with *njera* flatbread.

iv. Egusi soup (Nigeria) – *Egusi* refers to the seeds of a type of watermelon, ground up to make this popular soup, which also typically contains leafy vegetables, meat and fish.

v. Ndolé (Cameroon) – A stew made with *ndoleh*, a bitter leaf indigenous to West Africa. It also contains peanuts, beef and sometimes shrimp, served with *bobolo*, a loaf made from fermented cassava.

vi. Matoke (Uganda) – A banana indigenous to southwest Uganda is also the name of the country's national dish. The unripe fruit is steamed and mashed, then served with meat or veg.

vii. Piri-piri chicken (Mozambique) – Portuguese influences loom large in Mozambican cuisine, not least in this chicken dish with chilli, red pepper, garlic, vinegar and lemon – made globally popular by Nando's.

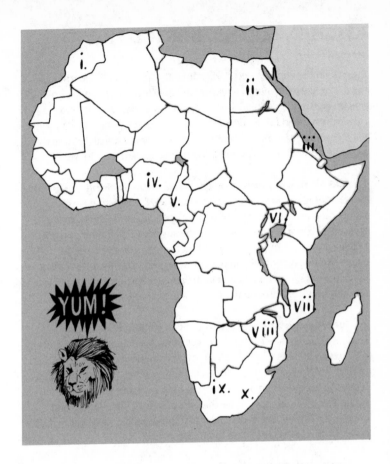

viii. Kapenta (Zimbabwe) – Otherwise known as the Tanganyika sardine, *kapenta* refers to two similar species of freshwater fish, which are eaten (both fresh and dried) with *sadza*, or maize porridge.

ix. Smileys (South Africa) – The name is uncannily apt: you get a big toothy grin from these sheeps' heads roasted whole over a *braai* in townships around South Africa.

x. Cape breyani (South Africa) – A descendent of the Persian *biryani*, this rice-and-meat combo crossed over from India with the Cape Malay community in the early days of the slave trade.

KITCHEN MYTHS, BUSTED

Even the most experienced chefs admit it: you never stop learning in the kitchen. Just as there's always some new technique to pick up and marvel at, there are also bits of accepted wisdom in your repertoire – rules you've been following without hesitation for years – that turn out to be completely unnecessary or, worse, counterproductive. More than a couple of the following debunkings came as a surprise to me.

1. Add oil to the pot to prevent pasta from sticking
While adding salt to your pasta water is good for boosting flavour, following it with a drizzle of olive oil, as people often advise, isn't a great idea. For one, the oil pools on the surface of the water and does little to prevent clumping. Then, when the pot is drained, the pasta gets coated in the layer of oil and becomes slippery, making it less capable of absorbing sauce. If you want to prevent sticking, simply boil a decent amount of water and give the pasta a stir after it goes in.

2. Sear a steak at the start to lock in its juices
This rule comes direct from the pages of Auguste Escoffier, one of the most revered chefs of all time, so it must be right, right? Wrong. Various people who know what they're up to in the kitchen, including J Kenji López-Alt at *Serious Eats*, have tested this repeatedly and found that steaks seared at the end rather than the start lose fewer juices and cook more evenly. The most important thing is to let the meat rest after cooking: this allows moisture to be reabsorbed.

3. Don't cook vegetables if you want maximum nutrition
The theory often put forward by raw food advocates is that cooking vegetables kills valuable nutrients. It's true that the nutritional composition can be altered by heat, but not always in a bad way. Some vegetables, such as carrots, courgettes and broccoli, are generally more nutritious boiled than raw. Of course there are trade-offs: the increase in carotenoid levels when you boil a carrot comes at the expense of falling polyphenols. The best advice is to enjoy your veg in a variety of ways.

4. Wait for food to cool before putting it in the fridge

Food is less likely to spoil if it cools slower – and hot food in the fridge will mess with the circulation of cold air. Neither claim stands up. Old-fashioned iceboxes might struggle to cope with the heat but not modern fridges (though they may expend more energy in the process). And food bacteria can double every 30–40 minutes, so it's better to refrigerate straight away.

5. Always serve pork well done

Fears about undercooking pork are not without basis: they mostly come down to trichinosis, a parasitic disease which humans can catch if the internal temperature of the meat doesn't go above 58°C/136°F during cooking. The good news is that trichinosis is much less common in pork these days due to higher processing standards. Salmonella is still a risk, but if you buy good-quality meat and cook it to 62°C/144°F (medium-rare) or above, you're very unlikely to have a problem.

6. Stir risotto constantly while adding stock

The traditional method is to add stock to the rice in small batches and stir constantly until it's done. This way, more starch is rubbed off the rice, leading to a creamier risotto – and more evenly cooked rice. The logic doesn't quite hold. If you cook the risotto at a low heat in a wide frying pan and stir it just once, as J Kenji López-Alt has demonstrated, you'll get risotto that's evenly cooked and no less creamy.

7. Put an avocado stone in guacamole to keep it from turning brown

Avocados (and bananas and apples) turn brown after you cut them open because they contain an enzyme called polyphenol oxidase, which oxidizes when exposed to air. The notion that an avocado stone in your guacamole will prevent this chemical process from happening is nonsense – it'll only work on the patch of guacamole directly underneath it. A better solution is to squeeze in some lime to delay oxidation (at no cost to the flavour) and seal out the air with clingfilm. Now turn to page 96 for a recipe courtesy of rock legend and guacamole fan Jack White.

BEEF UP!

Actors who gained weight for movie roles.

Jared Leto for *Chapter 27* (2007)
<u>Gained:</u> 67lb to play John Lennon's assassin
Mark Chapman.
<u>Method:</u> Microwaved pints of ice cream with olive oil
and soy sauce. "Really, it's a stupid thing to do," he
told the *Guardian*. "I got gout, and my cholesterol
went up so fast in such a short time that my doctors
wanted to put me on [heart disease drug] Lipitor."

Russell Crowe for *Body of Lies* (2008)
<u>Gained:</u> 63lb to play a CIA chief.
<u>Method:</u> "Bring on the burgers, baby," he said,
adding: "If you want to put on weight, you just elect
to live a sedentary lifestyle."

Matthew McConaughey for *Gold* (2016)
<u>Gained:</u> 40lb to play a modern-day prospector.
<u>Method:</u> "For six months... I was a real yes man," he
said. "Pizza night could be any night. Cheeseburger
and beer for breakfast was a great idea too. I really
relaxed on the rules."

**Renée Zellweger for *Bridget Jones's Diary*
(2001)**
<u>Gained:</u> 20lb – and had to do it again for the sequel
The Edge of Reason (2004)
<u>Method:</u> Big Macs and milkshakes for breakfast,
curries, pizzas and doughnuts for lunch, pasta and
steaks for dinner – and strictly no exercise.

Robert De Niro for *Raging Bull* (1980)
<u>Gained:</u> 60lb to play boxer Jake LaMotta in his
later years.
<u>Method:</u> Three large meals a day with an emphasis
on pasta, meat, butter, ice cream and beer. "The first
15lb was fun, the rest was hard work," he said.

SLIM DOWN!

Actors who lost weight for movie roles.

Christian Bale for *The Machinist* (2004)
<u>Lost:</u> 62lb to play a troubled factory worker.
<u>Method:</u> Lived on a cup of coffee and an apple or tin of tuna per day and smoked to curb his appetite, reportedly putting his health in grave danger in the process.

Ashton Kutcher for *Jobs* (2013)
<u>Lost:</u> 18lb to play Steve Jobs.
<u>Method:</u> Followed Jobs' fruitarian diet (*see* page 138) and was subsequently hospitalized with pancreas problems.

Matthew McConaughey for *Dallas Buyers Club* (2013)
<u>Lost:</u> 50lb to play AIDS patient and activist Ron Woodroof.
<u>Method:</u> Restricted himself to Diet Coke, egg whites and one piece of chicken per day, and chewed a lot of ice. "As soon as I hit 143lb," he said, "I started losing my eyesight."

Anne Hathaway for *Les Misérables* (2012)
<u>Lost:</u> 25lb to play Fantine.
<u>Method:</u> Dropped 10lb before shooting by adhering to a cleanse, then a further 15lb during production by eating just two thin pieces of dried oatmeal paste a day. "It was definitely a little nuts," she conceded.

50 Cent for *All Things Fall Apart* (2011)
<u>Lost:</u> 54lb to play a football player diagnosed with cancer.
<u>Method:</u> Went on a nine-week liquid diet and did three hours on a treadmill each day.

CHANGING TASTES

Global food demand is on the rise – by 2050, according to some estimates, it will be almost double what it was in 2005. Why? For one, the world's population is growing and may reach 9.7 billion by 2050; another big factor is rising incomes in developing countries, leading to higher demand for meat and other sources of protein. These trends account for some, but not all, of the following dietary shifts.

UP

China ~ Pork
Total consumption has increased five-fold since 1980. China now consumes half of the world's pigs, almost 500 million a year.

USA ~ Almonds
Americans consume ten times as many almonds today as they did in 1965. This is influenced in large part by reports of their nutritional value.

South Africa ~ Poultry
Per capita consumption increased by almost 80 per cent between 2000 and 2014 – poultry is a relatively cheap form of protein.

World ~ Soy
Global production expanded nearly ten-fold between 1961 and 2009. Approximately 75 per cent of soybeans are used for animal feed.

DOWN

USA ~ Whole milk
Per capita consumption has decreased by 78 per cent since 1970, with the health benefits of milk under scrutiny and dairy-free diets on the rise.

China ~ Sweet potatoes
Per capita consumption went down from 227g (8oz) per day in 1963 to 99g (3½oz) per day in 2003 – due, in part, to China's shift to a more protein-rich diet.

UK ~ White bread
Purchases have fallen by 75 per cent since 1974, while those of brown and wholemeal bread have risen by 85 per cent.

France ~ Wine
In 2010, 17 per cent of the population drank wine on a daily basis, down from 51 per cent in 1980. The shift from outdoor labour to office work is one factor, crackdowns on drink driving another.

THE FORGOTTEN QUEEN OF ICES

Mrs Beeton has cast a long shadow over British Victorian cookery, but there are other near-contemporary figures who deserve just as much attention. One is Eliza Acton, author of the brilliant *Modern Cookery for Private Families* (1845). Another is the author and entrepreneur **Mrs Marshall**, known in her time as the "Queen of Ices".

Born Agnes Bertha Smith in 1855, she was raised in Walthamstow and "practised at Paris and Vienna under celebrated chefs", according to the *Pall Mall Gazette*. In 1883, with her husband Alfred, she opened Marshall's School of Cookery in central London. More than just a school, it sold all kinds of kitchen equipment – including devices of Mrs Marshall's own invention. Particularly noteworthy is her Marshall's Patent Freezer, which claimed to produce "smooth and delicious" ice cream in just three minutes – an extraordinary boast considering that most modern machines take half an hour or more and require overnight freezing beforehand.

The claim has been tested on a surviving model and deemed more or less accurate (it's closer to five minutes). No less impressive are Mrs Marshall's four beautifully written cookbooks, including *The Book of Ices* (1885) and *Mrs AB Marshall's Book of Cookery* (1888). The school survived for several decades after her death in 1905, as did the weekly paper *The Table*, which she launched in 1886. Otherwise, her reputation has largely – and for no good reason – faded from memory.

UDDER

In his diary of 11 October 1660, Samuel Pepys wrote: "After we had done there Mr Creed and I to the Leg in King Street, to dinner, where he and I and my Will had a good udder". Boiled cow's udder (known as "elder" in Britain) was a fairly commonplace dish in Pepys's time but has been in decline ever since. Even lovers of tripe rarely encounter it, though udder is still occasionally seen for sale in West Yorkshire, as well as in Italy, Belgium and France, where it is known as *tétine de veau*.

THE MAN WHO LOVES ROADKILL

Since he was a teenager in the 1950s, **Arthur Boyt** has been eating animals and birds found dead on the roads of England. There are few things the retired Cornish marine biologist won't take home and pop in the Aga, from beached whales to badgers in a state of advanced decay. Controversy is rarely far away – his fondness for dogmeat won't win him many fans – though he would argue that roadkill is much more ethical than eating farmed meat, which he generally avoids. Here he tells me about some of his favourite roadside finds.

Rabbit

"I'm happiest to see a rabbit, because our cat only eats roadkill rabbits. They're very plentiful here in Cornwall. If the rabbit is badly mashed I'll throw it over the fence so the crows can eat it."

Dog

"Of all the roadkill I've eaten, the one I like best is dog. The meat is quite sweet and tender: I generally put a joint in the bottom of the Aga and let it simmer for two or three hours. I've only eaten three dogs: two lurchers and a beagle. Most dogs have a collar and you can't very well eat it if it's obvious there's an owner."

Badger

"I find a lot of badgers on the road. My favourite part is the head, because you get a range of different tastes and textures: the masseter muscle; the tongue, which is like any other tongue; and the salivary glands, which taste almost like soft roe."

Otter

"Otters are very nice! I would usually fry an otter with onions and garlic, then casserole it for three hours. I do occasionally cut it into strips and fry it, but I don't ever eat anything rare because I want to make sure it's properly dead."

Fox

"Now, you see, if you just skin and cook a fox, it pongs of fox. But I discovered an old Italian recipe where you skin it and throw it in a river for three days, and that takes away the smell. It tasted like lamb, it was delicious!"

Squirrel

"Often with a squirrel I'll throw it out on the moor, but if I cook it then I cut the legs off. If it's a male then I'll remove its balls and collect them until I have a good, umm, meal of squirrel balls."

Grass snake

"Once I found a grass snake in the London area, that was quite interesting. There wasn't much of the snake to eat, but there were very nice snake eggs on the road, and when I cooked them they tasted like egg yolks."

Sparrows

"As for birds, I would eat just about anything. In days gone by I would cycle 100 miles from Watford to Norwich and you could pick up half-a-dozen dead sparrows. I haven't seen a dead sparrow for years. We don't have them nesting around here anymore."

NOT ALL OF ARTHUR'S ROADKILL EXPERIMENTS HAVE TURNED OUT QUITE SO APPETIZING...

Cat

"I've eaten a few cats. They weren't very tasty, but perhaps the owners hadn't fed them enough. I'd marinate them in cider, that's one way of improving the flavour. It's the same as rat and squirrel: tender, good for you, but not a lot of taste in it."

Horseshoe bats

"I ate a horseshoe bat once. There's not much to it really."

VERJUICE

The unsweetened liquid of sour fruit such as crab-apples or unripe grapes, verjuice was much more popular in western cuisine during Roman and medieval times than it is today. Valued as a way of adding acidity to a wide variety of dishes, it was eventually superseded in the kitchen by lemon juice and vinegar. It deserves a comeback: as an ingredient and as a refreshing drink.

A BRIEF HISTORY
OF THE RESTAURANT

It's often said that the restaurant emerged out of the French Revolution, when cooks employed by the aristocracy found themselves jobless and began catering instead for the masses. This account has been challenged by recent historians who believe restaurants were operating in Paris 20 years before the revolution. And though many of the conventions that define modern restaurants were formed in Paris in the late 18th century, the history of eating out goes much deeper and wider than that.

Tabernae in Ancient Rome (from 5th century BC) – A *taberna* was a single-room shop where a variety of goods were sold, including wine and cooked food. They were mainly used by travellers and those who couldn't afford a private kitchen.

Song Dynasty eating houses (11th century AD) – In Kaifeng, the capital of China during the Song Dynasty (960–1279), there were hundreds of commercial food businesses patronized by rich and poor alike. Menus and table service were part of the offering.

Tipping (16th century) – The practice of giving gratuities for good service is often traced back to Tudor England. One dining establishment frequented later by Samuel Johnson (1709–84) had a bowl printed with the words "To Insure Promptitude" – and some have wondered if "tip" was an acronym for this phrase.

Coffeehouses (late 17th century) – Spreading from Mecca via the Ottoman Empire, the coffeehouse becomes popular in Europe as a place to meet, eat and drink.

The modern restaurant emerges (1760s) – Historians disagree on the exact moment of origin, but it's clear that new things were happening in the Paris dining scene in the late 18th century. Rather than taking what was on offer at a *table d'hôte*, you could now order from a selection of dishes, at a time of your choosing, and sit alone or with friends rather than at a communal table. Printed menus, uniformed waiters and other things we associate with modern restaurants soon became commonplace.

"Restaurateur" defined (1771) – The *Dictionnaire de Trévoux* explains the term as "someone who has the art of preparing true broths, known as 'restaurants', and the right to sell all kinds of custards, dishes of rice, vermicelli and macaroni, egg dishes" and so on. The word "restaurant" referring to an eating house, rather than a restorative broth, first appears in a decree of 1886.

French Revolution (1789–99) – The great upheaval in French society floods the nascent restaurant industry with chefs formerly employed in aristocratic kitchens.

Restaurants spread to America (1790s) – The French concept crosses the Atlantic, with new-style restaurants taking root in Boston and other cities. French restaurants remain synonymous with fine dining for centuries to come.

Service à la russe (mid 19th century) – The system of plating food in the kitchen and serving it to individual diners, rather than putting all items of a service on the table together (*service à la française*), becomes prevalent in Europe.

White Castle founded (1921) – America's first burger chain launches in Wichita, Kansas, laying the foundations for the modern fast-food restaurant. The franchise model developed by McDonald's in the 1950s enabled fast-food chains to expand much more rapidly.

Chefs' tables (late 20th century) – Chefs have been inviting friends and family to dine in restaurant kitchens for a long time, but over the past few decades it has been repackaged as a special privilege for paying guests.

Underground restaurants (early 21st century) – Also known as supper clubs, these private dining enterprises, often located in people's homes, have become extremely popular over the past decade.

Gourmet food trucks (early 21st century) – Street vendors have been wheeling food around for hundreds of years, but only this century have mobile food preparation units become associated with gourmet food, with American cities leading the trend. Roy Choi's Kogi in Los Angeles (established 2008) is a notable pioneer.

GREAT FOOD MARKETS OF THE WORLD

There are few pleasures greater than wandering into a busy food market in a foreign country and submitting to the great whirl of sensory impressions – spice colours and bird squawks, fruit tastes and fish smells – hitting you from every angle. I can vividly recall the extreme heat of a tiny chilli I ate as a teenager at the Blantyre market in Malawi, and the pungent intensity of Hue's Dong Ba market in Vietnam after a rainstorm, and the kaleidoscope of unfamiliar flavours at the Zapote farmers' market in San José, Costa Rica. Here are some other notable examples.

Vast – La Central de Abasto in Mexico City is the world's largest wholesale food market, sprawling over 328 hectares (810 acres) and housing more than 2,000 businesses, mostly fruit, veg and meat wholesalers. Each day, it disgorges 30,000 tons of food, fulfilling an estimated 80 per cent of the consumption needs of the Mexico City metropolitan area (some 25.4 million people). See also:

» Rungis in Paris is, at 234 hectares (578 acres), larger than Monaco.

Old – In 2014, Borough Market in London celebrated its 1,000th anniversary. The dating here is a bit sketchy, based as it is on a fleeting reference in an Icelandic historical account written 200 years later; also, the market moved around a bit before settling on its present site by London Bridge. Let's just say it's seen plenty of history, and that today, with scores of traders operating beneath the railway arches, it's still going strong. See also:

» La Boqueria in Barcelona traces its origins back to the 13th century.

Fishy – So vibrant is the Tsukiji fish market in central Tokyo that its status as a tourist attraction threatens to overwhelm its daily operation – shifting more than 2,000 tons of seafood every day. As a result, only 120 visitors are allowed into the daily tuna auctions. These take place in the inner market, which focuses on wholesale, while the retailers and restaurants – including rivals Sushi Dai and Daiwa-Zushi – do brisk business outside. See also:

» Noryangjin in Seoul has over 1,000 seafood varieties, many of them still alive.

Fruity – Azadpur Mandi in Delhi is said to be Asia's largest wholesale market for fruit and vegetables. Every day, thousands of city vendors flock here to stock up on pineapples and lemons, red onions and giant orange pumpkins – the range of produce is staggering – and business continues all through the night. See also:

» New Covent Garden Market in London, the largest wholesale fruit and veg market in the UK.

Spicy – At odds with the outsized modernity of Dubai, the Spice Souk in the east of the city is a reminder of a time before skyscrapers and megamalls. Down its narrow passages, you'll find colourful piles of turmeric, cinnamon, nutmeg, rose petals and dried fruit. Haggling may not go down so well at the Mall of the Emirates but it's very much expected here. See also:

» The Spice Bazaar in Istanbul, built in the 1660s, was the last stop for camel caravans travelling on the Silk Road from China.

FIVE MORE FOOD MARKETS TO VISIT

St Lawrence Market (Toronto, Canada)
Ver-o-Peso (Belém, Brazil)
Mercato il Capo (Palermo, Italy)
Or Tor Kor (Bangkok, Thailand)
Queen Victoria Market (Melbourne, Australia)

WACKAGING

Once, food and drink packaging simply stated its contents and storage instructions. Now it wants to engage you in a jokey chat. *Guardian* journalist Rebecca Nicholson coined a word for this in 2011 – "wackaging" – and blamed the trend on Innocent Drinks, the English juice and smoothie company which has been playing with its packaging copy since 1999. ("Separation occurs," warned one of its juice cartons, "but mummy still loves daddy.") Countless other products, from sandwiches to teabags, now come with wackaging as standard.

THE SIMPLEST CHEESECAKE

This cheesecake, made for us in May 2016 by the winemaker Stephanie Tscheppe-Eselböck of Gut Oggau in southeast Austria, has only four ingredients and is wondrously easy to prepare. You could complicate it by adding a biscuit base, grating some lemon rind into the mix, and/ or topping the finished cake with soft fruit before serving. But it went down a treat just as it was.[1]

Serves 6

4 eggs, beaten

500g (1lb 2oz) ricotta

250g (9oz) soured cream

50g (1¾oz) sugar

Preheat your oven to 180°C (350°F/ gas mark 4), and line a 23-cm (9-inch) springform tin with greaseproof paper. Combine the ingredients in a large bowl with a spoon or spatula and transfer them to the tin. Bake until the cake is nicely browned on top, 40–45 minutes. Leave to cool for at least 10 minutes. Dust with icing sugar and serve with a dollop of cream.

[1] A large intake of wine may have helped. A number of our interviews for *The Gannet* have taken a distinctly boozy turn; one, with a very generous sommelier in Stockholm, went on for 14½ hours and involved a list of beers and wines as long as my arm. At Gut Oggau, Stephanie and her husband Eduard insisted we try their entire range as well as various experimental blends down in their cellar. "Each one brims with attitude and verve," I later wrote, "but, unlike some natural wines, they are well-balanced and beautifully structured too. Leaving Gut Oggau at 3 in the afternoon, having sampled most of their range, we are neither of those things. But we are happy." I'll end this note by saying I've made the cheesecake a couple of times since, under conditions of greater sobriety, and it was just as delicious.

XNIPEC

A fiery Yucatan salsa made with habanero peppers, tomato, onion and sour orange juice, served as you would a typical salsa (but with extra caution). The name (pronounced shnee-pek) translates from Mayan as "dog's nose", a reference to how much you will sweat while eating it.

FOOD HOTSPOTS:
THE ARTISTS' HANGOUT

La Colombe d'Or, Provence – Where is the dividing line between hostelry and art gallery? It's hard to tell at this 1920s inn in Saint-Paul-de-Vence, in the hills above Nice, where the dining room alone contains works by Picasso, Matisse, Braque and Miró. They were all guests – along with Chagall, Léger and Calder – of the original owner Paul Roux, who had no background in art but who befriended many artists seeking refuge in the south of France at that time. A sign on the hotel's entrance read: "*Ici on loge à cheval, à pied ou en peinture*", which roughly translates as: "Here we lodge those on foot, on horseback or with paintings" – and many of the works that still grace the hotel today were given in exchange for free accommodation.

Max Fish, New York – Before it was a bar, this Lower East Side stalwart was an informal art gallery – and after securing its liquor license in 1989 it continued showing work and acting as a hub for the local art scene. Owner Ulli Rimkus made (and still does, at a new location on Orchard Street) a point of employing artists as bartenders: employees over the years have included Kiki Smith and Nan Goldin.

Paris Bar, Berlin – Since it was taken over by art-world scenesters in the 1970s, this raucous French restaurant in Charlottenburg has been a magnet for local and visiting artists – evidence of their custom lines every wall. Yoko Ono, Jeff Koons and Robert Rauschenberg dined here, as did Madonna, David Bowie and Iggy Pop. The great German painter Martin Kippenberger paid for his steak frites and beer with canvases, one of which was later sold by the owner for €2.5m to dig the restaurant out of debt.

MORE ART-WORLD MAGNETS

Els Quatre Gats, Barcelona – Picasso, Ramon Casas I Carbó, Antoni Gaudí

The French House, London – Francis Bacon, Augustus John, Lucian Freud

LARAGH STUART'S BEEF ROULADEN

The Gannet interviewed food producer Laragh Stuart at her cottage in Co Wicklow amid spells of piercing sunshine and apocalyptic Irish showers. Laragh made the perfect meal for uncertain weather: a rich, satisfying beef rouladen, inherited from her late father. She remembers him preparing the little rolls of beef on the kitchen table with a bottle of red wine open beside him. Eating it as the rain belted down outside, I could appreciate why Laragh finds this dish so immensely comforting.

Serves 5

600g (1lb 5oz) fillet or flank of beef (ask your butcher for very thin steaks)

Dijon mustard (about 75g/2½oz)

15 thin slices of pancetta or prosciutto (about 200g/7oz)

15 gherkins

Butter

Olive oil

200g (7oz) pearl onions, peeled

35g (1¼oz) flour

1 tbsp paprika

Salt and pepper

Rapeseed oil (or sunflower)

300g (10½oz) chestnut mushrooms (or any type you prefer)

A few sprigs of fresh thyme, finely chopped

2 bay leaves

2 tbsp organic vegetable stock (in 200ml/7fl oz of water)

300ml (10fl oz) red wine

Lay some clingfilm on a board and place a steak on it. Place some clingfilm on top of it and with a meat tenderizer proceed to pound it to your preferred thinness. I like it almost paper-thin. Repeat this until you have thinned out all your steaks. Cut them into strips wide enough to hold your gherkins.

Spread a generous amount of Dijon mustard on each steak and place a pancetta strip on top. Roll the steak tightly around a gherkin. Secure the roll with one or two cocktail sticks. Heat some butter and olive oil in a small skillet and add the pearl onions, cover with parchment paper and let simmer gently for about 20 minutes until soft. Set aside.

Combine the flour, paprika, salt and pepper in a large bowl and gently coat your beef rouladen.

In a large pot, heat some rapeseed oil, add the rouladen in batches and brown them on all sides.

Return all the rouladen to the pot and add the softened pearl onions, mushrooms, thyme, bay leaves, stock and red wine. Place a lid on the pot and cook for 1 hour on medium heat, then take the lid off and cook for a further hour. (Alternatively, you can put the pot in an oven with the lid on for 2 hours at about 120°C/250°F/ gas mark ½.)

Remove all the cocktail sticks and serve with parsnip and potato mash, and an orange and parsley gremolata.

THE REAL DEAL

If your area is famous for making a certain type of food or drink, be it a cheese, a sausage or a sparkling wine, you'd get understandably upset if other areas started flogging a similar product using the same name. That's why legal protections such as the EU's protected designation of origin (PDO) exist – though the limitations are often far from simple.

What?	Where can it be made?
Stilton	Derbyshire, Leicestershire and Nottinghamshire (though not in the village of Stilton itself, which is in Cambridgeshire)
Cognac	Six zones around the town of Cognac in southwest France: Grande Champagne, Petite Champagne, Borderies, Fins Bois, Bons Bois and Bois Ordinaire
Cornish pasties	Can be baked anywhere but must be assembled in Cornwall, England
Tequila	The central-western Mexican state of Jalisco – plus a few municipalities in other states – using blue agave plants
Vologda butter	The Vologda region of northwest Russia – the country's first ever PDO product
Thuringia bratwurst	At least 51 per cent of the ingredients have to originate in Thuringia, central Germany
Roquefort	Must be matured in the Combalou caves near Roquefort-sur-Soulzon in the Aveyron region of France, using only milk from Lacaune sheep
Champagne	A region within the province of Champagne in northeast France, though 40 villages in neighbouring départements are being added to meet global demand
Newcastle Brown Ale	Anywhere. The brewery secured protected status in 2000, but then moved across the river to Gateshead – and then to Tadcaster in Yorkshire – and had to ask the EU to lift the restriction

CULINARY TIME TRAVEL: CZARIST RUSSIA

In the first chapter of her extraordinary memoir *Mastering the Art of Soviet Cooking*, Russian-born food writer **Anya von Bremzen** sets about recreating a pre-revolutionary Czarist feast at her mother's tiny apartment in Queens, New York. "It was our cultural heritage, but it was like a fruit you can't reach," she told me. "We felt we had a right to know how it tasted." I asked her to revisit that meal, which took a staggering amount of effort to produce, and place it in the context of its time.

The period

"In Moscow, about ten years before the revolution, there was a time of great artistic flourishing called the Silver Age. This, I think, was when Russian cuisine really came to its greatest fruition. Unlike westernized St Petersburg, where they were eating foie gras and oysters, Moscow retained its Russian identity. There were a lot of merchants from Siberia with a lot of money to burn, and taverns called *traktir* were serving very decadent à la russe fare. A few years later, it would all come crashing down."

Zakuski (appetizers)

"The whole reason for appetizers is to chase shots of vodka. At my mum's apartment, we served them on a separate table – that's how the guests would mingle. You'd have salty things: blini, herring, smoked salmon, smoked sturgeon and all kinds of pickles – wild mushroom, cucumber…"

Soup

"*Botvinya* is one of the dishes that completely disappears after the revolution. It's a chilled soup made with the fermented beverage *kvass*, to which you add smoked fish (sturgeon or salmon), carrots and green vegetables. Basically the idea is to put a crunchy salad inside a cold soup. To be honest, it was kind of weird. Maybe we didn't make it right, or maybe we've lost that flavour – everything was back then kind of fizzy and fermented."

Main course

"Then we made a very elaborate fish pie called *kulebiaka*. You can have it stuffed with cabbage and meat, but the really decadent one is with fish, wild mushrooms and layers of dilled rice. Between those layers you would put blini, which made absolutely no sense until we tried it – the blini absorbed the juices and separated the different fillings. This dish is extremely hard to make and impossible to find nowadays – we had only read about it in books."

Dessert

"We made *Guriev kasha*, a completely anachronistic dessert named after an early-19th-century Russian minister of finance. You make a sweetened semolina *kasha* and layer it with candied nuts and berries, and *penki*, which are the skins that form on cream when it's baked. I made my *penki* late at night and they spattered all over my oven."

To drink

"Russian food is generally very vodka-friendly, but it depends on the person. Myself, I would stick with vodka. In the old *traktir* (taverns), you might have Champagne with the soup. A lot of women would switch to wine – red or white, depending on the course – while the men would go on drinking vodka. This being Russia, people would get drunk and start smashing glasses, but the restaurants would maintain a sense of propriety. They would have been stratified: a little provincial family in one room, rich people in another. They knew how to keep the class situation under control."

IF YOU LOVE...

Let's start with the obvious: there's no such thing as the world's best wine or chocolate or cheese shop. But there *are* shops you could recommend to people who love wine or chocolate or cheese and know for sure they wouldn't be disappointed. Between us, the *Gannet* team have visited – and can unreservedly recommend – the following places.

CHOCOLATE ~ Chocolátl (Amsterdam, Holland)
What's exceptional about this shop in Jordaan is its range: it stocks bars by great chocolate-makers all over the world, including Dandelion from San Francisco and Rózsavölgyi from Budapest.

WINE ~ Chambers Street Wines (New York, USA)
A Lower Manhattan gem with an independent approach and a terrific selection, prioritizing European wines from less obvious regions. The free in-store tastings add to the appeal.

CHEESE ~ Laurent Dubois (Paris, France)
Restaurant critic and Paris expert Alec Lobrano raved to us about Laurent Dubois, who has three shops in the city, and we weren't disappointed. It's hard to imagine a better range of French cheeses.

TINNED FISH ~ Conserveira de Lisboa (Lisbon, Portugal)
Even if you dislike fish, this emporium in the Baixa district is worth visiting for the 1930s interior alone, its shelves stacked high with brightly coloured cans (they stock 130 varieties).

VINEGAR ~ Gegenbauer (Vienna, Austria)
Interviewing Erwin Gegenbauer for *The Gannet* in 2016, I sampled his whole range of vinegars and oils and could understand why top chefs like Thomas Keller and Alain Ducasse rated them so highly.

LIQUORICE ~ Kado (Berlin, Germany)
Salty liquorice is an acquired taste, but those partial to it cannot fail to love this shop in Kreuzberg, which sells sweets from 10 countries alongside liquorice teas, powders and toothpaste.

MUSTARD ~ Tierenteyn-Verlent (Ghent, Belgium)
Part of the appeal of this old-timer on the Groentenmarkt is that you can't buy their (excellent) mustard elsewhere. It's freshly made downstairs and poured into stoneware jars from a barrel in the corner.

BREAD ~ Tartine Bakery (San Francisco, USA)
Chad Robertson, who opened Tartine with his partner Elizabeth Prueitt in 2002, is now a bona fide bread god, and bakers from all over make pilgrimages to his original bakery and the new 5,000-square-foot Tartine Manufactory, both in the Mission district.

FRUIT ~ Sembikiya (Tokyo, Japan)
Perhaps the best way to approach this insanely expensive fruit boutique is to think of it as a gallery where you can simply marvel at the perfect specimens and their price tags: $237 for a single mango.

COFFEE ~ Sant'Eustachio Il Caffè (Rome, Italy)
New-wave coffee aficionados may scoff, but this old-school espresso bar and shop next to the Pantheon exudes everything that's great about Italian coffee culture: the atmosphere, the aroma, the attitude.

COOKBOOKS ~ Books for Cooks (London, UK)
The wonderful collection of old and new cookbooks (and other books about food) is half the story at this Notting Hill stalwart. There's also a tiny café at the back which picks out a book (or books) each day and cooks recipes from them.

YUBA

"Yuba is to tofu what burrata is to supermarket mozzarella," trumpeted Julia Kramer for *Bon Appetit* magazine in 2014. And yet few outside China or Japan have heard of yuba: the thin skin that forms on the surface of soy milk when it's heated to make tofu. Nutty-tasting and elastic, it can be rolled up and eaten fresh, or dried and sliced into noodles, or simmered in a broth as you would a piece of fish. Given its extraordinarily high levels of protein, it's no surprise that yuba is a popular meat alternative in East-Asian cooking.

NATIVE INGREDIENTS
OF AUSTRALASIA

i. Kutjera (Australia) – The caramel-like "desert raisin" is used in condiments, or ground to a powder and mixed into bread dough.

ii. Mulga apple (Australia) – The name is misleading: this is an insect gall (or tumour) growing in the wood of the mulga tree. It has a sweet white wasp grub in the centre, which is said to taste like apples. Commonly eaten by aborigines of Central Australia.

iii. Wattleseed (Australia) – Traditionally used by aborigines to make bush bread, these edible seeds have recently become popular as a flavouring – with coffee and hazelnut notes – for desserts, bread or beer.

iv. Lemon Myrtle (Australia) – The leaf of this flowering plant has a higher citral purity than lemongrass. It can be paired with fish, used to season desserts or made into a tea.

v. Snowberry (Tasmania) – Eaten raw by aborigines, these sour, summer-fruiting berries were a popular ingredient in colonial pies.

vi. Ambarella (Polynesia/Micronesia) – An egg-sized green fruit with crunchy flesh that turns yellow as it ripens. Flavour is reminiscent of pineapple and mango.

vii. Breadfruit (Melanesia) – The name refers to the texture and aroma of its cooked ripe flesh, similar to freshly baked bread. Related to the jackfruit. Seldom eaten raw. Sometimes mashed with coconut milk.

viii. Kava (Vanuatu) – A root used to produce a drink with an unappealing muddy taste but some interesting properties: it's used as a sedative and a pain reliever and is sometimes drunk instead of alcohol.

ix. Pikopiko (New Zealand) – Also known as "bush asparagus", this is the tip of an edible fern, such as the hen and chicken species. Often employed as a garnish to impart a forest flavour.

x. Common pipi (New Zealand) – An important part of the Maori diet, these edible clams are usually barbecued or eaten raw.

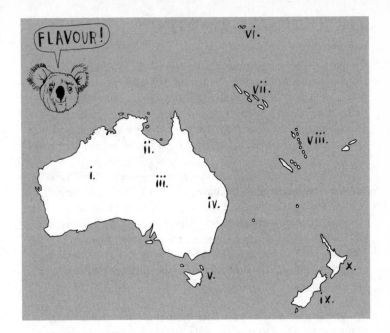

Z

ZOOPHAGY

The OED defines it as "The act or practice of eating animals or animal matter", though zoophagy is usually understood to mean the consumption of exotic animals. This practice was big in the 19th century when Europeans would return from the colonies with tastes for far-off fauna, which some at home were also curious to experience. Noted zoophagists included William and Frank Buckland (*see* page 13) and Peter Lund Simmonds, whose 1859 book *The Curiosities of Food* is an extraordinary compendium of global eating habits.

THE ULTIMATE MARTINI?

In what is probably the most famous drink order of all time, in Ian Fleming's novel *Dr No* (1958), James Bond asks for his Martini to be "shaken and not stirred". He also specifies Russian or Polish vodka, rather than gin, and a slice of lemon peel. It's a controversial request. Though classic Martinis call for just two or three ingredients, opinions diverge as to how exactly they should be served. Here are a choice few.

» At Duke's Hotel in London, where Fleming used to drink, head barman **Alessandro Palazzi** makes a Vesper Martini in the writer's honour. The glass and the vodka (Potocki) are frozen in advance. He pours three small drops of Angostura bitters into the frozen glass, then pours 20ml (¾fl oz) of amber vermouth on top, followed by 30ml (1fl oz) of vodka and 50ml (1 ¾fl oz) of gin (No 3) – no shaking or stirring involved. The drink is served garnished with the zest of an unwaxed orange.

» The Spanish film director **Luis Buñuel** recommended putting all the ingredients – glasses, gin and shaker – in the refrigerator the day before. "Don't take anything out until your friends arrive; then pour a few drops of Noilly Prat and half a demitasse spoon of Angostura bitters over the ice. Shake it, then pour it out, keeping only the ice, which retains a faint taste of both. Then pour straight gin over the ice, shake it again, and serve."[1]

» According to **Winston Churchill**, the only way to make a Martini was with ice-cold gin and a bow in the direction of France. **Alfred Hitchcock** took a similar approach, saying that the closest he wanted to get to a bottle of vermouth was looking at it from across the room.

» The food writer **Julia Child** preferred "reverse Martinis": a glass full of vermouth on the rocks with a topper of gin.

[1] Buñuel went on: "The making of a dry Martini should resemble the Immaculate Conception, for, as Saint Thomas Aquinas once noted, the generative power of the Holy Ghost pierced the Virgin's hymen 'like a ray of sunlight through a window' – leaving it unbroken."

FAMOUS LAST MEALS

History records a lot of weird details about those who loom large in its annals. It's thanks to the hard work of biographers, historians, journalists and autopsy reporters that we know what various notable figures ate just before they met their end. Based on the ten final meals below and the date of consumption, can you identify the diners in question?

1   Four scoops of ice cream and six chocolate chip cookies
(15 August 1977)

2   Goats' milk, cooked vegetables and oranges. A drink of ginger, sour lemons and strained butter mixed with aloe juice
(30 January 1948)

3   Coffee, orange juice, two boiled (five-minute) eggs, toast and marmalade on the side (22 November 1963)

4   Mock turtle soup, roast Virginia fowl with chestnut stuffing, baked yams and cauliflower with cheese sauce (14 April 1865)

5   Boiled chicken and rice, hot water laced with honey
(30 December 2006)

6   A slice of apple pie and a glass of milk (30 September 1955)

7   Honeyed cakes, Madeira wine, black bread (30 December 1916)

8   Mushroom and asparagus omelette, Dover sole and vegetable tempura (30 August 1997)

9   Corned beef sandwich (8 December 1980)

10 A bowl of French onion soup (13 August 2004)

1) Elvis Presley 2) Mahatma Gandhi 3) JFK 4) Abraham Lincoln 5) Saddam Hussein 6) James Dean 7) Rasputin 8) Princess Diana 9) John Lennon 10) Julia Child

ACKNOWLEDGEMENTS

This book has been in many ways a collective effort. For their research, fact-checking, proofing and infectious enthusiasm, my thanks to James Hansen and Ebony-Renee Baker. Adam Park helped make the book possible and pushed it along with unflagging energy, while Yousef Eldin and Dan Dennison fed in valuable suggestions on layout and design. Thanks to them and everyone else who has contributed time, energy and brainpower to The Gannet over the last few years.

Rob Fox helped me grasp how the book ought to work and was a great reader and reassurer, as were Ursula and Tony Fox, who got me interested in food in the first place (despite some early resistance on my part). Leo Nabarro, Emile Dinneen, Tom Tracey, James Coffey, Macey Marini, Ray O'Meara, Dan Stafford, Mark O'Connell, Ben Hofer, Sophie Dollar, Hugo Meyer Esquerré, Alban de Pury and Sophie Missing all offered sage advice and encouragement, and I'm particularly grateful to Karl Toomey for the creative chats.

Without Jon Elek, our lightning-quick agent, this book would not exist – and thanks also to Kat Aitken and Millie Hoskins at United Agents for all their brilliant work. Stephanie Jackson has been a constant source of support and inspiration and her team at Octopus is the best there is. Thanks to Pauline Bache for her sharp and sensitive editing, to Juliette Norsworthy for tip-top art direction, and to Ellen Bashford, Matthew Grindon and Caroline Brown for getting this book out into the world.

Diana Henry, Jay Rayner, Jeremy Lee, Dave Broom and Jeff Gordinier, your kind words are very much appreciated. And thank you to everyone who invited The Gannet into their homes, let us nose around their cupboards and fridges, and consented for their quotes and recipes to be used in this book.

Finally, my thanks and love to Emma Bennett who put up with late-night writing sessions, gave brilliant notes and said all the right things at the right moments. I can't imagine a better dining companion in the world than you.